Smart Innovation, Systems and Technologies

Volume 103

Series editors

Robert James Howlett, Bournemouth University and KES International,
Shoreham-by-sea, UK
e-mail: rjhowlett@kesinternational.org

Lakhmi C. Jain, University of Technology Sydney, Broadway, Australia;
University of Canberra, Canberra, Australia; KES International, UK
e-mail: jainlakhmi@gmail.com; jainlc2002@yahoo.co.uk

The Smart Innovation, Systems and Technologies book series encompasses the topics of knowledge, intelligence, innovation and sustainability. The aim of the series is to make available a platform for the publication of books on all aspects of single and multi-disciplinary research on these themes in order to make the latest results available in a readily-accessible form. Volumes on interdisciplinary research combining two or more of these areas is particularly sought.

The series covers systems and paradigms that employ knowledge and intelligence in a broad sense. Its scope is systems having embedded knowledge and intelligence, which may be applied to the solution of world problems in industry, the environment and the community. It also focusses on the knowledge-transfer methodologies and innovation strategies employed to make this happen effectively. The combination of intelligent systems tools and a broad range of applications introduces a need for a synergy of disciplines from science, technology, business and the humanities. The series will include conference proceedings, edited collections, monographs, hand-books, reference books, and other relevant types of book in areas of science and technology where smart systems and technologies can offer innovative solutions.

High quality content is an essential feature for all book proposals accepted for the series. It is expected that editors of all accepted volumes will ensure that contributions are subjected to an appropriate level of reviewing process and adhere to KES quality principles.

More information about this series at http://www.springer.com/series/8767

Anna Esposito · Marcos Faundez-Zanuy
Francesco Carlo Morabito · Eros Pasero
Editors

Quantifying and Processing Biomedical and Behavioral Signals

 Springer

Editors
Anna Esposito
Dipartimento di Psicologia
Università della Campania "Luigi
  Vanvitelli"
Caserta, Italy

and

International Institute for Advanced
  Scientific Studies (IIASS)
Vietri sul Mare, Italy

Marcos Faundez-Zanuy
Fundació Tecnocampus
Pompeu Fabra University
Mataro, Barcelona
Spain

Francesco Carlo Morabito
Department of Civil, Environmental,
  Energy, and Material Engineering
University Mediterranea of Reggio Calabria
Reggio Calabria, Italy

Eros Pasero
Dipartimento di Elettronica e
  Telecomunicazioni, Laboratorio di
  Neuronica
Politecnico di Torino
Turin, Italy

ISSN 2190-3018   ISSN 2190-3026   (electronic)
Smart Innovation, Systems and Technologies
ISBN 978-3-030-06976-6   ISBN 978-3-319-95095-2   (eBook)
https://doi.org/10.1007/978-3-319-95095-2

Preface

The proposed book posits its basis on an interdisciplinary research having as primary objectives to study aspects and dynamics of human multimodal signal exchanges and pattern recognition in medicine. The goal is to seek for invariant features through the cross-modal analysis of verbal and nonverbal interactional modalities in order to define the relative mathematical models and pattern recognition algorithms for implementing emotionally interactive cognitive architectures capable of performing believable actions when reacting to human user requests. The analysis will stem from realistic scenarios where human interaction is on the stage and aims to:

- Identifying new methods for data processing and data flow coordination through synchronization, temporal organization, and optimization of new encoding features (identified through behavioral analyses) combining contextually enacted communicative signals;
- Developing shared digital data repositories and annotation standards for benchmarking the algorithmic feasibility and the successive implementation of believable HCI systems.

Given the multidisciplinary character of the book, scientific contributes are from computer science, physics, psychology, statistics, mathematics, electrical engineering, and communication science. The contributions in the book cover different scientific areas, even though these areas are closely connected in the themes they afford and all provide fundamental insights for cross-fertilization of different disciplines. In particular, most of the chapters contributing to this book were first discussed at the international workshop on neural networks (WIRN 2017) held in Vietri sul Mare from June 14 to 16, 2017, in the special session on: "**Dynamics of signal exchanges**" organized by Anna Esposito, Antonietta M. Esposito, Sara Invitto, Nadia Mammone, Gennaro Cordasco, Mauro Maldonato, and Francesco Carlo Morabito; and the special session on "**Neural networks and pattern recognition in medicine**" organized by Giansalvo Cirrincione, Vitoantonio Bevilacqua.

The contributors to this volume are leading authorities in their respective fields. We are grateful to them for accepting our invitation and making (through their participation) the book a worthwhile effort. We owe deep gratitude to the Springer project coordinator for books production Mr. Ayyasamy Gowrishankar, the Springer executive editor Dr. Thomas Ditzinger, and the editor assistant Mr. Holger Schaepe, for their outstanding support and availability. The editors in chief of the Springer series Smart Innovation, Systems and Technologies, Profs. Jain Lakhmi C. and Howlett Robert James, are deeply appreciated for supporting our initiative and giving credit to our efforts.

Caserta, Itlay   Anna Esposito
Mataro, Spain   Marcos Faundez-Zanuy
Reggio Calabria, Italy   Francesco Carlo Morabito
Turin, Italy   Eros Pasero

Organization Committee

The chapters submitted to this book have been carefully reviewed by the following technical committee to which the editors are extremely grateful.

Technical Reviewer Committee

Altilio Rosa, Università di Roma "La Sapienza"
Alonso-Martinez Carlos, Universitat Pompeu Fabra
Angiulli Giovanni, Università Mediterranea di Reggio Calabria
Bevilacqua Vitoantonio, Politecnico di Bari
Bramanti Alessia, ISASI-CNR "Eduardo Caianiello" Messina
Brandenburger Jens, VDEh-Betriebsforschungsinstitut GmbH, BFI, Dusseldorf
Buonanno Amedeo, Università degli Studi della Campania "Luigi Vanvitelli"
Camastra Francesco, Università Napoli Parthenope
Carcangiu Sara, University of Cagliari
Campolo Maurizio, Università degli Studi Mediterranea Reggio Calabria
Capuano Vincenzo, Seconda Università di Napoli
Cauteruccio Francesco, Università degli Studi della Calabria
Celotto Emilio, Ca' Foscari University of Venice
Ciaramella Angelo, Università Napoli Parthenope
Ciccarelli Valentina, Università di Roma "La Sapienza"
Cirrincione Giansalvo, UPJV
Colla Valentina, Scuola Superiore S. Anna
Comajuncosas Andreu, Universitat Pompeu Fabra
Commimiello Danilo, Università di Roma "La Sapienza"
Committeri Giorgia, Università di Chieti
Cordasco Gennaro, Seconda Università di Napoli

De Carlo Domenico, Università Mediterranea di Reggio Calabria
De Felice Domenico, Università Napoli Parthenope
Dell'Orco Silvia, Università degli Studi della Basilicata
Diaz Moises, Universidad del Atlántico Medio
Droghini Diego, Università Politecnica delle Marche
Ellero Andrea, Ca' Foscari University of Venice
Esposito Anna, Università degli Studi della Campania "Luigi Vanvitelli" and IIASS
Esposito Antonietta Maria, Sezione di Napoli Osservatorio Vesuviano
Esposito Francesco, Università di Napoli Parthenope
Esposito Marilena, International Institute for Advanced Scientific Studies (IIASS)
Faundez-Zanuy Marcos, Universitat Pompeu Fabra
Ferretti Paola, Ca' Foscari University of Venice
Gallicchio Claudio, University of Pisa
Giove Silvio, University of Venice
Giribone Pier Giuseppe, Banca Carige, Financial Engineering and Pricing
Kumar Rahul, University of South Pacific
Ieracitano Cosimo, Università degli Studi Mediterranea Reggio Calabria
Inuso Giuseppina, University Mediterranea of Reggio Calabria
Invitto Sara, Università del Salento
La Foresta Fabio, Università degli Studi Mediterranea Reggio Calabria
Lenori Stefano, University of Rome "La Sapienza"
Lo Giudice Paolo University "Mediterranea" of Reggio Calabria
Lupelli Ivan, Culham Centre for Fusion Energy
Maldonato Mauro, Università di Napoli "Federico II"
Manghisi Vito, Politecnico di Bari
Mammone Nadia, IRCCS Centro Neurolesi Bonino Pulejo, Messina
Maratea Antonio, Università Napoli Parthenope
Marcolin Federica, Politecnico di Torino
Martinez Olalla Rafael, Universidad Politécnica de Madrid
Matarazzo Olimpia, Seconda Università di Napoli
Mekyska Jiri, Brno University
Micheli Alessio, University of Pisa
Militello Carmelo, Consiglio Nazionale delle Ricerche (IBFM-CNR), Cefalù (PA)
Militello Fulvio, Culham Centre for Fusion Energy
Monda Vincenzo, Università degli Studi della Campania "Luigi Vanvitelli"
Morabito Francesco Carlo, Università Mediterranea di Reggio Calabria
Nardone Davide, Università di Napoli "Parthenope"
Narejo Sanam, Politecnico di Torino
Neffelli Marco, University of Genova
Parisi Raffaele, Università di Roma "La Sapienza"
Paschero Maurizio, University of Rome "La Sapienza"
Pedrelli Luca, University of Pisa
Portero-Tresserra Marta, Universitat Pompeu Fabra
Principi Emanuele, Università Politecnica delle Marche
Josep Roure, Universitat Pompeu Fabra

Rovetta Stefano, Università di Genova (IT)
Rundo Leonardo, Università degli Studi di Milano-Bicocca
Salvi Giampiero, KTH, Sweden
Sappey-Marinier Dominique, Université de Lyon
Scardapane Simone, Università di Roma "La Sapienza"
Scarpiniti Michele, Università di Roma "La Sapienza"
Senese Vincenzo Paolo, Seconda Università di Napoli
Sesa-Nogueras Enric, Universitat Pompeu Fabra
Sgrò Annalisa, Università Mediterranea di Reggio Calabria
Staiano Antonino, Università Napoli Parthenope
Stamile Claudio, Université de Lyon
Statue-Villar Antonio, Universitat Pompeu Fabra
Suchacka Grażyna, Opole University
Taisch Marco, Politecnico di Milano
Terracina Giorgio, Università della Calabria
Theoharatos Christos, Computer Vision Systems, IRIDA Labs S.A.
Troncone Alda, Seconda Università di Napoli
Vitabile Salvatore, Università degli Studi di Palermo
Xavier Font-Aragones, Universitat Pompeu Fabra
Uncini Aurelio, Università di Roma "La Sapienza"
Ursino Domenico, Università Mediterranea di Reggio Calabria
Vasquez Juan Camilo, University of Antioquia
Vesperini Fabio, Università Politecnica delle Marche
Vitabile Salvatore, Università degli Studi di Palermo
Wesenberg Kjaer Troels, Zealand University Hospital
Walkden Nick, Culham Centre for Fusion Energy
Zucco Gesualdo, Università di Padova

Sponsoring Institutions

International Institute for Advanced Scientific Studies (IIASS) of Vietri S/M (Italy)
Department of Psychology, Università degli Studi della Campania "Luigi Vanvitelli" (Italy)
Provincia di Salerno (Italy)
Comune di Vietri sul Mare, Salerno (Italy)
International Neural Network Society (INNS)
Università Mediterranea di Reggio Calabria (Italy)

Contents

Part I
Introduction

Chapter 1
A Human-Centered Behavioral Informatics

Anna Esposito, Marcos Faundez-Zanuy, Francesco Carlo Morabito and Eros Pasero

Abstract Currently, researchers coming from psychological, computational, and engineering research fields have developed a human-centered behavioral informatics characterized by techniques analyzing and coding human behaviors, conventional and unconventional social conducts, signals coming from audio and video recordings, auditory and visual pathways, neural waves, neurological and cognitive disorders, psychological and personal traits, emotional states, mood disorders. This interweaving of expertise had produced extensive research progresses and unexpected converging interests allowing the groundwork for a book dedicated to pose the current progresses in dynamics of signal exchanges and reporting the latest advances on the synthesis and automatic recognition of human interactional behaviors. Key features considered are the fusion and implementation of automatic processes and algorithms for interpreting, tracking, and synthesizing dynamic signals such as facial expressions, gaits, EEGs, brain and speech waves. The acquisition, analysis, and modeling of such signals is crucial for computational studies devoted to a human-centered behavioral informatics.

A. Esposito (✉)
Dipartimento di Psicologia, Università della Campania "Luigi Vanvitelli",
Caserta, Italy
e-mail: iiass.annaesp@tin.it

A. Esposito
IIASS, Vietri Sul Mare, Italy

M. Faundez-Zanuy
Pompeu Fabra University, Barcelona, Spain
e-mail: faundez@tecnocampus.cat

F. C. Morabito
Università degli Studi "Mediterranea" di Reggio Calabria, Reggio Calabria, Italy
e-mail: morabito@unirc.it

E. Pasero
Dip. Elettronica e Telecomunicazioni, Politecnico di Torino, Turin, Italy
e-mail: eros.pasero@polito.it

© Springer International Publishing AG, part of Springer Nature 2019
A. Esposito et al. (eds.), *Quantifying and Processing Biomedical and Behavioral Signals*, Smart Innovation, Systems and Technologies 103,
https://doi.org/10.1007/978-3-319-95095-2_1

Keywords User modelling · Artificial intelligence · Customer care · Daily life
activities · Biometric data · Social signal processing · Social behavior and
context · Complex Human-Computer interfaces

1.1 Introduction

When it comes to the implementation of socially and believable cognitive systems,
the multimodal facets of socially and emotionally colored interactions has been
neglected. Only partial aspects have been accounted for, such as the implementation
of automatic systems separately devoted either to identify vocal or facial emotional
expressions (a review is presented in [3, 5]). This is because at the date, there are no
studies that holistically approached the analysis of emotionally colored interactional
exchanges integrating speech, faces, gaze, head, arms, and body movements in a
unique percept in order to disclose the underlying cognitive abilities that merge
information from different sensory systems. This ability links cognitive processes
that converge and intertwine at the functional level and cannot understood if studied
separately and completely disentangled from one another [12].

"*The definition and comprehension of this link can be understood only identifying
some meta-entities describing brain processes more complex than those devoted to
the simple peripheral pre-processing of the received signals*" (see [6, p. 2]). Here,
the concept of meta-entities is intended to describe how humans merge at the same
time both symbolic and sub-symbolic knowledge in order to act and be successful in
their everyday living. To implement a human-centered informatics, there is a need
to understood these facets. There is a need of behavioral analyses performed on
multimodal communicative signals instantiated in different scenarios and automated
through audio/video processing, synthesis, detection, and recognition algorithms [2,
4]. This will allow to mathematically structure in a unique unit both symbolic and
sub-symbolic concepts in order to define computational models able to understand
and synthesize the human ability to rule individual choices, perception, and actions.
There is a need to inform computation on how human-centered behavioral systems
can be implemented.

The methodological approach is simple and quite easy to state and is summarized
along the three macro-analyses reported below:

- Identifying distinctive and invariant interactional features describing mood states,
  social conducts, beliefs, and experiences through the cross modal analysis of
  audio and video recordings of multimodal interactional signals (speech, gestures,
  dynamic facial expressions, and gaze)
- Defining the encoding procedures that mathematically describe such features and
  assessing their robustness both perceptually (on the human side) and of computationally (on the automatic side)

- Developing new algorithms for the extraction (trough automatic processing), detection, recognition, of such features together the delineating of new mathematical models of emotionally and socially colored human-machine interactions.

These easy to state macro-analyses are however, hard to be implemented. The cross-modal analysis of audio and video recordings requires specific experimental set-ups describing emotional and socially believable interactions and the selection of appropriate scenarios. The identification of possible scenarios will drive and contextualize the generation of the experimental data. For example, if the selected scenario considers the interaction among friends or relatives, the collected data will miss to provide data on interactional behaviors adopted among first encounters. A different scenario must be selected to collect data on such interactional exchanges. The questions here are how many scenarios must be considered? How many data must be collected? Why humans do not need all of it and adapt to any specific situations? Why we cannot provide machine with a human automaton level of intelligence?

These will remain open questions as long as we will not understood the struggling of the human endeavor. This reasoning can be set a part in an experimental context where an attempt is made to model a crumb of humanity. More practical questions are the following: Should participants wear appropriate sensors, cyber-gloves, as well as, facial and body markers to allow the collection of biofeedback data? How many cameras must be employed to have a full control of the interactional exchange? Should the cameras provide calibrated stereo full-body video of participants and/or a close-up of the participant's head to capture head, gaze, and facial features? In addition, appropriate data format and annotation standards must be selected to make available the digital data repositories to the scientific community. The dialogues and elicitations techniques may be re-defined iteratively in accord to the need raised by the experimental and algorithmic validation. This data will provide realistic models of human-human interaction in defined scenarios in order to assess project benchmarks that consider an agent, avatars, robot, or any socially believable ICT interface interacting with humans in similar scenarios

The assessment of the features robustness and the identification their relative encoding procedures requires the transcription of the data, the envisaging of possible data representation, and the evaluation of the amount of the interactional information captured by the selected data representation. This will produce qualitative and quantitative features assigned to the multimodal signals produced during the interactions. The qualitative analysis is generally first performed by an expert and then by naïve judges. The transcription and the encoding modalities provided by the expert judges will serve as a reference guide for the assessment requested to the naïve ones. The qualitative assessment must be performed separately on each signal (audio, video, gestures, and facial expressions) as well as on the multimodal combination in order to evaluate the amount of emotional/social information conveyed by the single and combined communication modes. The quantitative assessment must be implemented by ICT experts through the automatic extraction of acoustic and video features, such as F0; energy; linear prediction coefficients; hand motions; facial marker motions, exploiting standard signal processing techniques redefined through qualitative analysis offered by the expert and naïve judges [1, 8, 10, 13, 14]. Biometric data maybe included such as heath rate, EEGs, and more [7, 9, 11].

Finally, the feature correlations should appropriately align altogether the multi-modal interactional signals and manually label them according to their symbolic and sub-symbolic informational content in order to build up a mathematical model of the interactional exchanges. Data processing and data flow coordination algorithms will be applied to the labeled data in order to synchronize and temporally organize the encoding features identified through behavioral analyses. Correlations among features and their contributions in conveying meaningful/emotional information will be identified in order to structure the corresponding meta-entities (mathematical concepts) exploiting both the symbolic and sub-symbolic information gathered. Statistical analyses will assess their significance, and eliminate redundant information in order to reduce data dimensions and consequently the computational costs associated with an algorithmic exploitation.

The above-described procedures will produce information on the dynamic of signal exchanges, the one that is needed for a human-centered informatics and for quantifying and processing biomedical and behavioral signals.

1.2   Content of This Book

The themes tackled in this book are related to the most recent efforts for quantifying and processing biomedical and behavioral signals. The content of the book is organized in sections, each dedicated to a specific topic, and including peer-reviewed, not published elsewhere, chapters reporting applications and/or research results in quantifying and processing behavioral and biomedical signals. The seminal content of the chapters was discussed for the first time at the International Workshop on Neural Networks (WIRN 2017) held in Vietri sul Mare, Italy, from the 14th to the 16th of June 2017. The workshop, being at its 28th edition is nowadays a historical and traditional scientific event gathering together researcher from Europe and overseas.

Section I introduces methods and procedures for quantifying and processing biomedical and behavioral signals through a short chapter proposed by Esposito and colleagues.

Section II is dedicated to the dynamics of signal exchanges. It includes 19 short chapters dealing with different expertise in analyzing, processing, and interpreting behavioral and biometric data. Of particular interest are analyses of visual, written and audio information and corresponding computational efforts to automatically detect and interpret their semantic and pragmatic contents. Related applications of these interdisciplinary facets are ICT interfaces able to detect health and affective states of their users, interpret their psychological and behavioral patterns and support them through positively designed interventions.

Section III is dedicated to the exploitation of neural networks and pattern recognition techniques in medicine. This section includes 5 short chapters on computer-assisted approaches for clinical and diagnostic diseases. These advanced assistive technologies are currently at a pivotal stage, even though they keep the promise to improve the quality of life of their end users and facilitate medical diagnoses.

1.3 Conclusion

To date, there are no studies that combine the analysis of speech, gestures, facial expressions and gaze to holistically approach human socially and emotionally colored interactions. Studies collecting data of such interactions would be considered a scientific basis for the validation of automatic algorithms and cognitive prototypes and will constitute a new benchmark to be used in quantifying and processing dynamics of signal exchanges among humans. The analysis of this data will serve the scientific community at the large and contribute to the implementation of cognitive architectures which incorporate and integrate principles from psychology, biology, social sciences, and neurosciences, and demonstrate their viability through validation in a range of challenging social, autonomous assistive benchmark case studies. These architectures will be able to gather information and meanings in the course of everyday activity. They will build knowledge and show practical ability to render the world sensible and interpretable while interacting with users. They will be able to understand interactional behavioural sequences that characterize relevant actions for collaborative learning, sharing of semantic and pragmatic contents, decision making and problem solving

Acknowledgements The research leading to the results presented in this paper has been conducted in the project EMPATHIC (Grant No: 769872) that received funding from the European Union's Horizon 2020 research and innovation programme.

References

1. Atassi, H., Esposito, A., Smekal, Z.: Analysis of high-level features for vocal emotion recognition. In: Proceedings of 34th IEEE International Conference on TSP, pp. 361–366, Budapest, Hungary (2011)
2. Esposito, A., Jain, L.C.: Modeling social signals and contexts in robotic socially believable behaving systems. In: Esposito A., Jain L.C. (eds.) Toward Robotic Socially Believable Behaving Systems Volume II - "Modeling Social Signals", ISRL vol. 106, pp. 5–13. Springer International Publishing Switzerland, Basel (2016)
3. Esposito, A., Esposito, A.M., Vogel, C.: Needs and challenges in human computer interaction for processing social emotional information. Pattern Recogn. Lett. **66**, 41–51 (2015)
4. Esposito, A., Palumbo, D., Troncone, A.: The influence of the attachment style on the decoding accuracy of emotional vocal expressions. Cogn. Comput. **6**(4), 699–707 (2014). https://doi.org/10.1007/s12559-014-9292-x
5. Esposito, A., Esposito, A.M.: On the recognition of emotional vocal expressions: motivations for an holistic approach. Cogn. Process. J. **13**(2), 541–550 (2012)

6. Esposito, A: COST 2102: Cross-modal analysis of verbal and nonverbal communication (CAVeNC). In: Esposito A. et al. (eds.) Verbal and Nonverbal Communication Behaviours. LNCS, vol. 4775, pp. 1–10. Springer International Publishing Switzerland, Basel (2007)

7. Faundez-Zanuy, M., Hussain, A., Mekyska, J., Sesa-Nogueras, E., Monte-Moreno, E., Esposito, A., Chetouani, M., Garre-Olmo, J., Abel, A., Smekal, Z., Lopez-de-Ipinã, K.: Biometric applications related to human beings: there is life beyond security. Cogn. Computat. **5**, 136–151 (2013)

8. Justo, R., Torres, M.I.: Integration of complex language models in ASR and LU systems. Pattern Anal. Appl. **18**(3), 493–505 (2015)

9. Mammone, N., De Salvo, S., Ieracitano, C., Marino, S., Marra, A., Corallo, F., Morabito, F.C.: A permutation disalignment index-based complex network approach to evaluate longitudinal changes in brain-electrical connectivity. ENTROPY **19**(10), 548 (2017)

10. Mohammadi, G., Vinciarelli, A.: Automatic personality perception: prediction of trait attribution based on prosodic features. IEEE Trans. Affect. Comput. **3**(3), 273–283 (2012)

11. Morabito, F.C., Campolo, M., Labate, D., Morabito, G., Bonanno, L., Bramanti, A., de Salvo, S., Marra, A., Bramanti, P.: A longitudinal EEG study of Alzheimer's disease progression based on a complex network approach. Int. J. Neural Syst. **25**(2) (2015). https://doi.org/10.1142/S0 129065715500057

12. Maldonato, M., Dell'Orco, S., Esposito, A.: The emergence of creativity. World Future: J. New Paradigm Res. **72**(7–8), 319–326 (2016)

13. Prinosil, J., Smekal, Z., Esposito, A.: Combining features for recognizing emotional facial expressions in static images. In: Esposito A. et al. (eds.) Verbal and Non-verbal Features of Human-Human and Human-Machine Interaction, LNAI, vol. 5042, pp. 56–69. Springer, Berlin (2008)

14. Vinciarelli, A., Esposito, A., André, E., Bonin, F., Chetouani, M., Cohn, J.F., Cristan, M., Fuhrmann, F., Gilmartin, E., Hammal, Z., Heylen, D., Kaiser, R., Koutsombogera, M., Potamianos, A., Renals, S., Riccardi, G., Salah, A.A.: Open challenges in modelling, analysis and synthesis of human behaviour in human–human and human–machine interactions. Cogn. Comput. **7**(4), 397–413 (2015)

Part II
Dynamics of Signal Exchanges

Chapter 2
Wearable Devices for Self-enhancement and Improvement of Plasticity: Effects on Neurocognitive Efficiency

Michela Balconi ⓘ **and Davide Crivelli** ⓘ

Abstract Neurocognitive self-enhancement can be defined as a voluntary attempt to improve one's own cognitive skills and performance by means of neuroscience techniques able to influence the activity of neural structures and neural networks sub-serving such skills and performance. In the last years, the strive to improve personal potential and efficiency of cognitive functioning lead to the revival of mental train-ing activities. Recently, it has been suggested that such practices may benefit from the support of mobile computing applications and wearable body-sensing devices. Besides discussing such topics, we report preliminary results of a project aimed at investigating the potential for cognitive-affective enhancement of a technology-mediated mental training intervention supported by a novel brain-sensing wearable device. Modulation of motivational and affective measures, neuropsychological and cognitive performances, and both electrophysiological and autonomic reactivity have been tested by dividing participants into an experimental and an active control group and by comparing the outcome of their psychometric, neuropsychological, and instru-mental assessment before, halfway through, and after the end of the intervention period. The technology-mediated intervention seemed to help optimizing attention regulation, control and focusing skills, as marked by a reduction of response times at challenging computerized cognitive tasks and by the enhancement of event-related electrophysiological deflections marking early attention orientation and cognitive control. Available evidences, together with the first set of findings here reported, are starting to consistently show the potential of available methods and technologies for enhancing human cognitive abilities and improving efficiency of cognitive processes.

Keywords Cognitive enhancement · Wearable device · Mindfulness
Neurofeedback · Mobile computing

M. Balconi · D. Crivelli
Research Unit in Affective and Social Neuroscience, Catholic University of the Sacred Heart, 20123 Milan, Italy
e-mail: davide.crivelli@unicatt.it

M. Balconi (✉) · D. Crivelli
Department of Psychology, Catholic University of the Sacred Heart, 20123 Milan, Italy
e-mail: michela.balconi@unicatt.it

© Springer International Publishing AG, part of Springer Nature 2019
A. Esposito et al. (eds.), *Quantifying and Processing Biomedical and Behavioral Signals*, Smart Innovation, Systems and Technologies 103,
https://doi.org/10.1007/978-3-319-95095-2_2

2.1 Neurocognitive Enhancement: Novel Perspectives

The growing complexity and competitiveness in society and professional contexts and the drive to ever greater performances fuelled the debate on the potential and the opportunities of different methods and techniques capable to enhance cognitive abilities, thought ethical implication of such applications are subject of hot debates [1].

Neurocognitive self-enhancement can be defined as a voluntary attempt to improve one's own cognitive skills and behavioural performance by means of neuroscience techniques able to influence the activity of neural structures and neural networks subserving such skills and supporting cognitive performance. At the basis of neurocognitive enhancement is the idea that empowerment of cognitive abilities and neural efficiency can be empowered across all the lifespan by systematic activation and re-activation of cortical-subcortical networks mediating cognitive functions, thus fostering brain plasticity—understood as the ability of neural structures to strengthen existing connections and create new ones based on experience and training.

Different neuroscience tools and techniques have been classically used to promote brain plasticity and to help neural systems to strengthen and optimise their connections, the mostly investigated ones being non-invasive brain stimulation and neurofeedback techniques. While the former are based on externally-induced stimulation or modulation of ongoing neural activity and do not necessarily require the active engagement of the stimulated individual, the latter critically grounds on the active role of the participant since they apply the principles of operant conditioning and, thus, promote plasticity and cognitive empowerment by training participants' self-awareness and active control over physiological correlates of cognitive skills [2]. It has been suggested that such peculiar feature of neurofeedback empowerment interventions might have additional effects on long-term retention of training effects, since the participants is directly involved in finding and consolidating personalized strategies to intentionally modulate their neurophysiological activity.

In recent years, the strive to improve personal potential and efficiency of cognitive functioning also lead to the revival and the renewed diffusion of mental training activities. Indeed, a growing literature on the effects of mental training and meditation practice highlighted its potential for modulating overt behaviour and covert psychophysiological activity [3, 4] and for inducing short-term and—likely—long-term empowerment effects on cognitive and emotion regulation skills [5, 6]. In particular, mindfulness practice has attracted much attention and its application for self-empowerment in non-clinical settings notably grew, likely because it allows the practiser to train focusing, monitoring and attention skills by engaging and maintaining a specific aware and attentive mindset [5, 7, 8]. In Western culture, mindfulness is defined a peculiar form of mental training based on self-observation and awareness practices focused on the present and requiring conscious *intentional* focusing on and acceptance of one's own bodily sensations, mental states, and feelings, *nonjudgementally*, and *moment by moment* [9].

Neuroimaging and electrophysiological investigations allowed for defining neurofunctional correlates of such practice and for testing their effect with respect to neural activation patterns and biomarkers of neural activity. Going down to specifics, mindfulness practice has been associated to the activation of intrinsic functional connections and of a broad fronto-parietal network that includes cortical structures involved in the development of some key functions such as the definition of the self, planning, problem solving and emotional regulation [10].

Again, observed increase of medial and dorsolateral prefrontal cortex activity associated to mindfulness practice may mirror the up-regulation of cortical mechanisms mediating emotion regulation and self-monitoring [11–13]. Furthermore, the advantage in terms of attention orientation and information-processing efficiency provided by meditation trainings [14, 15] might be linked to the down-modulation of the activity of the default mode network often observed during meditation [16] and, in meditators, even at rest [17, 18]. Such reduced organized activity in large-scale default mode network has been associated to a decrease in self-referential mind-wandering and to a more efficient allocation of resources towards specific targets of attention. Interestingly, it was shown that during meditation—as compared to resting—even functional connectivity within the default mode network is reduced, and such reduction grows with mediation expertise [17, 18]. Consistently, it was also recently reported that even brief interventions based on mindfulness mental training (duration: 2 weeks) may induce the reorganization of the functional connectivity of large-scale brain networks involved in attention, cognitive and affective processing, awareness and sensory integration, and reward processing—which include, among other structures: bilateral superior occipital/middle gyrus, bilateral frontal operculum, bilateral superior temporal gyrus, right superior temporal pole, bilateral insula, caudate and cerebellum [19].

A similar picture of increased monitoring and attention regulation correlates is sketched even by electrophysiological research. Namely, meditation mental training have been often associated to the increase of alpha and theta activity over frontal areas [8], which may mark the progressive modulation of attention resources. Furthermore, the effects of such training practice on brain functioning have been supported even by finer-grained investigations based on event-related potentials recording—an electrophysiological technique that can inform on the progression of information-processing steps with excellent time resolution. In particular, the increase in neural efficiency and optimization of attention orienting processes seems to be almost systematically marked by the modulation of N2 and P3 event-related potentials [20–22], which are associated to attention regulation, allocation of cognitive resources, cognitive control and detection of novel relevant stimuli.

Reported effects of mindfulness practice with regard to the management of affective reactions and the improvement of cognitive functioning strengthen the considerations on its potential for promoting psychological well-being and enhancing cognitive abilities in non-clinical contexts. Nonetheless, such potential is hampered by a couple of methodological requirements of the mindfulness approach. Indeed, traditional mindfulness protocols (as well as meditation protocols overall) do require rather intense exercise and constant commitment. Such requirements limit the accessibility to meditation practice, and often lead to a gradual decrease of motivation and, consequently, to the suspension of individual practices [6, 23–25]. The impact of such limitations might be lowered thanks to the support of external devices and dedicated apps capable to reduce the demand of practice and to reward people by showing them clearly their progresses over time [6, 24]. In particular, motivation to practice and keep practising is usually fostered by allowing them to share their experience in dedicated communities, by proving practisers with easily-understandable and captivating graphs base on individual data, and by structuring their activities via goal-setting, milestones and rewards.

Besides promoting motivation and rewarding constant practice, such mobile computing applications might also contribute to subjective and objective outcomes of meditation activity by making practisers more aware of their advancements and level of engagement. Consistently, we recently proposed that accessibility of information to aware processing—in line with Cleeremans's theorization of human consciousness and models of the access consciousness [26]—is a crucial trigger for learning and for self-enhancement [6].

Along the same line, adaptive changes in neurocognitive activity and related changes in brain connectivity induced by mental training might be further helped by providing practisers with additional valuable information on the modulation of their psychophysical states due to practice.

Wearable devices and user-friendly non-invasive sensors capable to track and quantify physiological arousal and neural activity do provide, to date, actual opportunities to make practisers access implicit markers of internal bodily states and process such information reflectively and consciously. Indeed, implicit information of bodily states can be deemed as a kind of pre-conscious data, which are primarily unaware but can enter the spotlight of consciousness thanks to top-down attentional [27] or higher-level monitoring and meta-cognitive mechanisms [28]. According to the neural global workspace theory [27, 29], metacognitive and attention mechanisms may, in particular, exert their influence by amplifying medium-range resonant neural loops that maintain the representation of the stimulus temporarily active in a sensory buffer but still outside awareness. Such amplification might ignite global reverberating information exchanges supported by wide brain activation and long-distance connections between perceptual and associative cortical areas, which are thought to connote conscious processing of mental contents.

Notwithstanding the potential of integrated wearable devices and mobile computing solutions with regard to mental training and self-enhancement interventions, available scientific literature on such topic is still mainly constituted by reports on technology-mediated protocols based on smartphone/tablet apps, with only limited data on the contribution of non-invasive body-sensing devices. In the next section, we will then present preliminary data on the outcome and efficacy of a mental training protocol supported by a non-invasive brain-sensing device with regard to neurocognitive efficiency, with the additional aim to foster the debate on novel integrated mind-brain empowerment techniques.

2.2   Combining Mental Training and Wearable Brain-Sensing Devices: An Applies Example

2.2.1   Project Aims

The primary aims of the project were to investigate the potential for cognitive enhancement and improvement of stress management of an intensive technology-mediated mental training intervention, where training practice was supported by a novel commercial wearable device—the Lowdown Focus glasses (SmithOptics Inc, Clearfield, UT, USA)—capable of sensing brain activity and informing the user on changes of their electrophysiological profiles in real-time. Besides testing behavioral, cognitive, autonomic, and electrophysiological outcomes of the technology-mediated intervention, the project also aimed at exploring subjective perceived efficacy and at validating the usability of the device. Here we will on preliminary findings on neurocognitive efficiency modulations induced by the intervention.

2.2.2   Sample

A total of 38 volunteer participants ($M_{age} = 23.58$, $SD_{age} = 1.92$; $M_{edu} = 17.15$, $SD_{edu} = 1.31$) were enrolled. Criteria for inclusion were: age range 20–30 yo; mild-moderate stress levels; normal or corrected-to-normal hearing and vision. History of psychiatric or neurological diseases, presence of cognitive deficits, ongoing concurrent therapies based on psychoactive drugs that can alter central nervous system functioning, clinically relevant stress levels, occurrence of significant stressful life events during the last 6 months, and preceding systematic meditation experience were instead exclusion criteria. All participants signed a written informed consent to participate in the project. All procedures and techniques followed the principles of the Declaration of Helsinki and were reviewed and approved by the Ethics Committee of the Department of Psychology of the Catholic University of the Sacred Heart.

2.2.3   Procedure

Participants were divided into an experimental and an active control group and underwent preliminary, intermediate and post-intervention assessments before, halfway and after the end of the intervention period. Figure 2.1 represents the overall structure of the project and its steps.

The assessment procedure includes three levels of measurement: psychometric measures related to motivational, affective and personality profile; neuropsychological and behavioral measures related to cognitive performance; and instrumental

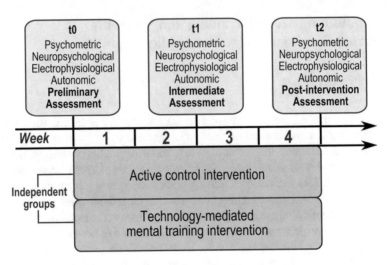

Fig. 2.1 Overall structure and main steps of the project

measures (EEG-ERPs, autonomic indices) to assess individual electrophysiological and autonomic profiles at rest and during cognitive stress. Psychometric measures included data on anxiety and stress levels, coping and stress management abilities, motivational-affective traits, personality profiles, and mindfulness-related, self-observation and bodily awareness skills. Neuropsychological and behavioral measures included data on problem-solving, attention, short-term memory, focus and executive control abilities. Finally, instrumental measures included electrophysiological and autonomic indices of bodily activity at rest (eyes closed and eyes open) and during an activating cognitive task. As for EEG, planned analyses of resting-state and task-related data included both frequency-domain and time-domain indices, so to investigate potential modulations of the oscillatory profile or information-processing markers induced by the training. During both resting and the activating task, participants' autonomic activity was also monitored and recorded, so to track potential physiological arousal modulations and individual autonomic regulation ability in different conditions (at rest and with respect to a cognitive stressor).

Both experimental and active control training procedures requested participants to plan and complete daily sessions of practice for four weeks. Participants' commitment has been systematically manipulated so that it gradually increased across the weeks, from 10 min a day at the beginning of the intervention to 20 min a day during the latest sessions. Such critical aspect of the training was devised to keep challenging participants and to continue stimulating their progresses and skills. The experimental training was based on the use of the Lowdown Focus wearable device—namely, a pair of glasses with an embedded EEG-neurofeedback system connected to a dedicated smartphone app that was devised to support mental training and mindfulness meditation practices. Awareness-promoting activities primarily hinge on mental focusing, sustained self-monitoring, and intentionally paying attention to breathing and related bodily sensations. Similarly, the active control group underwent a control intervention that was comparable in its overall structure, in the amount of commitment it required, and in the modalities of fruition with respect to the experimental one, apart from two critical aspects of the experimental training: participants' active agent role and the support of the Lowdown Focus device in providing real-time feedbacks on participants' mental states. All participants were also requested to be constant in their practices, to systematically plan training activities at the same moment of the day, and to keep track of the hour when they actually run the training sessions, so to try and control for potential influence of circadian rhythms on cognitive and physiological processes [6].

2.2.4 Results and Discussion

Changes in affective-motivational and perceived stress levels, cognitive and behavioral performance, and metrics of electrophysiological and autonomic activity induced by the trainings was qualitatively and quantitatively identified by comparing preliminary, intermediate and post-intervention assessment data. In particular, here we briefly present a first set of data concerning participants' performance at two challenging computerized tasks tapping on attention regulation and cognitive control skills, as well as first data concerning an early event-related electrophysiological deflection (namely the N2 component) marking attention orienting and response control. To control for potential biases due to inter-individual differences, reported analyses were performed on weighted modulation indices, which were computed by rationalizing halfway and final values over baseline values for each of the above-described outcome measures. Across-group statistical comparisons were performed via independent-groups t-tests (PASW Statistics 18, SPSS Inc, Quarry Bay, HK).

In addition to standardized and widely-used paper-and-pencil neuropsychological tests, efficiency of information-processing and cognitive skills have been assessed via computerized testing. Participants had to complete both a Stroop-like task (Stim2 software, Compumedics Neuroscan, Charlotte, NC, USA)—where colour-word stimuli were randomly and rapidly presented on a screen and the examinee had to signal whether the word and the colour in which it is written were congruent (e.g. the word "RED" written in red) or incongruent (e.g. the word "BLUE" written in green)—and a standardized battery of reaction times tasks that tap on different aspect of attention functions, from simple alertness and vigilance to higher cognitive control and response inhibition mechanisms—the MIDA battery [30]. While the active control group did not show any relevant modulation of their performance at the Stroop-like task, the experimental group already presented a significantly greater reduction of response times after two weeks of practice ($t_{(32)} = -3.157, p = .003$), and an even greater reduction at the end of the intervention ($t_{(28)} = -2.658, p = .013$). Figure 2.2 reports groups' performances at the Stroop-like task. Such first findings suggest that participants undergoing the experimental technology-mediated intervention showed an increasing optimization of information-processing and responses,

Fig. 2.2 Modulation of participants' performance at a computerized Stroop-like task (percentage changes weighted on individual baseline performance). Bars mirror mean changes of response times (RT) for the Active Control (light grey) and Experimental (dark grey) groups, halfway through the interventions (w-t1) and at the end of the intervention (w-t2). Error-bars represent ±1 SE

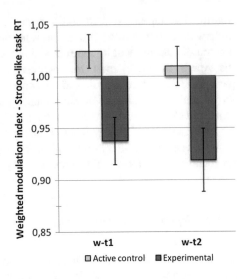

which may mirror increased cognitive efficiency—i.e. more timely and still accurate responses to visual stimuli.

The hypothesis of increased cognitive efficiency is corroborated even by data concerning the performance at the MIDA battery. Indeed, during the most complex and effortful subtask of the battery—a form of Go/No-go task devised to stress behavioural inhibition and executive control skills—the participants who underwent the experimental intervention showed significantly reduced reaction times with respect to the active control group at the end of the training ($t_{(29)} = -3.340, p = .002$). Figure 2.3 reports groups' performances at the complex subtask of the MIDA battery. Again, the reduction of reaction times together with very limited errors and uncontrolled responses suggest that, while executing the task, participants were focused and succeeded in efficiently allocating attention resources to optimize stimulus-response patterns.

Furthermore, preliminary analyses of task-related electrophysiological data highlighted consistent modulations of early attention-related markers. Task-related EEG data were recorded during the computerized Stroop-like task via a 16-channel system (V-Amp system, Brain Products GmbH, Gilching, Germany; SR = 1000 Hz) and then filtered, cleaned, checked for artifacts, and processed offline to compute event-related potentials (ERP). Event-related potentials are small deflections of scalp electrical potential that can be used to study different steps of a cognitive process. The amplitude of the N2 ERP, in particular, has been used as a marker of attention regulation, allocation of attention resources, and cognitive control mechanisms [31]. Present data highlighted that the amplitude of the N2 component over frontal areas notably increased in the experimental group at the end of the training, while it essentially remained at baseline levels in the active control group (significant between-group difference: $t_{(26)} = 2.147, p = .041$). Figure 2.4 reports the modulation of N2 amplitude for the experimental and the active control groups. Such up-modulation of N2

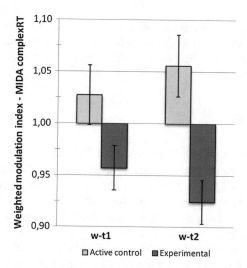

Fig. 2.3 Modulation of participants' performance at the complex reaction times subtask of the standardized MIDA battery (percentage changes weighted on individual baseline performance). Bars mirror mean changes of response times (RT) for the Active Control (light grey) and Experimental (dark grey) groups, halfway through the interventions (w-t1) and at the end of the intervention (w-t2). Error-bars represent ±1 SE

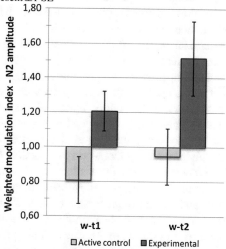

Fig. 2.4 Modulation of participants' N2 event-related potential recorded in response to target stimuli during a computerized Stroop-like task (percentage changes weighted on individual baseline measures). Bars mirror mean changes of peak amplitude for the Active Control (light grey) and Experimental (dark grey) groups, halfway through the interventions (w-t1) and at the end of the intervention (w-t2). Error-bars represent ±1 SE

amplitude is in line with previous evidences on the effects of long-term meditation [8, 32] and—given the functional interpretation of such ERP component [31]—hints at the presence of an enhancement of attention orientation and executive focus.

2.3 Conclusions

Available evidences, together with the first set of findings here reported, are starting to consistently show the potential of available methods and technologies for enhancing human cognitive abilities and improving efficiency of cognitive processes. Namely, we observed that four weeks of intensive mental training based on mindfulness principles and, *critically*, supported by non-invasive, highly-usable brain-sensing wearable device helped practisers to train and optimize the efficiency of attention regulation, control and focusing skills, as marked by a reduction of response times with no concurrent loss of accuracy at challenging computerized cognitive tasks and by the enhancement of event-related electrophysiological deflections marking early attention orientation and cognitive control. We still have to acknowledge that scientific literature on such potential is still scarce or often limited to uncontrolled testing. Further properly-designed empirical investigations and replication studies are needed to sketch a reliable and broader picture of the contribution of technology-mediated training intervention to human neurocognitive enhancement.

Building on available data on the effect of mental training and meditation on brain structure, connectivity, and activity [19, 20, 33]—we nonetheless suggest that present set of behavioral and electrophysiological findings begin to coherently define a scenario of increased neurocognitive efficiency. Such scenario of optimized cognitive resources allocation and, consequently, optimized performance may reflect plastic changes in neural connectivity, which lead to better information-exchange in the broad frontal-parietal network mediating vigilance, attention, and monitoring processes [34]. In line with the neural global workspace model [27, 29] and hypothesis on implicit learning processes [6, 26], the availability of valuable information on internal mental states to mental training practisers thanks to the support of the wearable brain-sensing device and the ease of processing such additional information flow thanks to the supporting smartphone application had likely fostered strengthening of trained focusing and self-monitoring skills. Such training effect likely acted by enhancing the potential amplification effect of attention and monitoring skills that is crucial for igniting global reverberating information exchanges between neural structures, which is supported by connections between perceptual and associative cortical areas and which mediate higher cognitive abilities. Future research should better explore such possibility via structural and functional investigation tools—such as, effective connectivity analyses based on imaging and electrophysiological data and paired pulse transcranial magnetic stimulation—to properly qualify and quantify dynamical changes in brain information exchange induced by technology-mediated mental training.

References

1. Farah, M.J., Illes, J., Cook-Deegan, R., Gardner, H., Kandel, E., King, P., Parens, E., Sahakian, B., Wolpe, P.R.: Neurocognitive enhancement: what can we do and what should we do? Nat. Rev. Neurosci. **5**, 421–425 (2004). https://doi.org/10.1038/nrn1390
2. Enriquez-Geppert, S., Huster, R.J., Herrmann, C.S.: Boosting brain functions: improving executive functions with behavioral training, neurostimulation, and neurofeedback. Int. J. Psychophysiol. **88**, 1–16 (2013). https://doi.org/10.1016/j.ijpsycho.2013.02.001
3. Pascoe, M.C., Thompson, D.R., Jenkins, Z.M., Ski, C.F.: Mindfulness mediates the physiological markers of stress: systematic review and meta-analysis. J. Psychiatr. Res. **95**, 156–178 (2017). https://doi.org/10.1016/j.jpsychires.2017.08.004
4. Quaglia, J.T., Braun, S.E., Freeman, S.P., McDaniel, M.A., Brown, K.W.: Meta-analytic evidence for effects of mindfulness training on dimensions of self-reported dispositional mindfulness. Psychol. Assess. **28**, 803–818 (2016). https://doi.org/10.1037/pas0000268
5. Keng, S.-L., Smoski, M.J., Robins, C.J.: Effects of mindfulness on psychological health: a review of empirical studies. Clin. Psychol. Rev. **31**, 1041–1056 (2011). https://doi.org/10.1016/j.cpr.2011.04.006
6. Balconi, M., Fronda, G., Venturella, I., Crivelli, D.: Conscious, pre-conscious and unconscious mechanisms in emotional behaviour. Some applications to the mindfulness approach with wearable devices. Appl. Sci. **7**, 1280 (2017). https://doi.org/10.3390/app7121280
7. Khoury, B., Lecomte, T., Fortin, G., Masse, M., Therien, P., Bouchard, V., Chapleau, M.-A., Paquin, K., Hofmann, S.G.: Mindfulness-based therapy: a comprehensive meta-analysis. Clin. Psychol. Rev. **33**, 763–771 (2013). https://doi.org/10.1016/j.cpr.2013.05.005
8. Cahn, B.R., Polich, J.: Meditation states and traits: EEG, ERP, and neuroimaging studies. Psychol. Bull. **132**, 180–211 (2006). https://doi.org/10.1037/0033-2909.132.2.180
9. Kabat-Zinn, J.: Full catastrophe living: using the wisdom of your body and mind to face stress, pain, and illness. Bantam Dell, New York (1990)
10. Raichle, M.E.: The brain's default mode network. Annu. Rev. Neurosci. **38**, 433–447 (2015). https://doi.org/10.1146/annurev-neuro-071013-014030
11. Adolphs, R.: The social brain: neural basis of social knowledge. Annu. Rev. Psychol. **60**, 693–716 (2009). https://doi.org/10.1146/annurev.psych.60.110707.163514
12. Phan, K.L., Wager, T., Taylor, S.F., Liberzon, I.: Functional neuroanatomy of emotion: a meta-analysis of emotion activation studies in PET and fMRI. Neuroimage **16**, 331–348 (2002). https://doi.org/10.1006/nimg.2002.1087
13. Bush, G., Luu, P., Posner, M.I.: Cognitive and emotional influences in anterior cingulate cortex. Trends Cogn. Sci. **4**, 215–222 (2000)
14. Jha, A.P., Krompinger, J., Baime, M.J.: Mindfulness training modifies subsystems of attention. Cogn. Affect. Behav. Neurosci. **7**, 109–119 (2007). https://doi.org/10.3758/CABN.7.2.109
15. Slagter, H.A., Lutz, A., Greischar, L.L., Nieuwenhuis, S., Davidson, R.J.: Theta phase synchrony and conscious target perception: impact of intensive mental training. J. Cogn. Neurosci. **21**, 1536–1549 (2009). https://doi.org/10.1162/jocn.2009.21125
16. Tomasino, B., Fregona, S., Skrap, M., Fabbro, F.: Meditation-related activations are modulated by the practices needed to obtain it and by the expertise: an ALE meta-analysis study. Front. Hum. Neurosci. **6**, 346 (2013). https://doi.org/10.3389/fnhum.2012.00346
17. Berkovich-Ohana, A., Harel, M., Hahamy, A., Arieli, A., Malach, R.: Alterations in task-induced activity and resting-state fluctuations in visual and DMN areas revealed in long-term meditators. Neuroimage **135**, 125–134 (2016). https://doi.org/10.1016/j.neuroimage.2016.04.024
18. Berkovich-Ohana, A., Harel, M., Hahamy, A., Arieli, A., Malach, R.: Data for default network reduced functional connectivity in meditators, negatively correlated with meditation expertise. Data Brief **8**, 910–914 (2016). https://doi.org/10.1016/j.dib.2016.07.015
19. Tang, Y.-Y., Tang, Y., Tang, R., Lewis-Peacock, J.A.: Brief mental training reorganizes large-scale brain networks. Front. Syst. Neurosci. **11**, 6 (2017). https://doi.org/10.3389/fnsys.2017.00006

20. Lutz, A., Slagter, H.A., Dunne, J.D., Davidson, R.J.: Attention regulation and monitoring in meditation. Trends Cogn. Sci. **12**, 163–169 (2008). https://doi.org/10.1016/j.tics.2008.01.005

21. Malinowski, P., Moore, A.W., Mead, B.R., Gruber, T.: Mindful aging: the effects of regular brief mindfulness practice on electrophysiological markers of cognitive and affective processing in older adults. Mindfulness (N. Y.) **8**, 78–94 (2017). https://doi.org/10.1007/s12671-015-0482-8

22. Moore, A., Gruber, T., Derose, J., Malinowski, P.: Regular, brief mindfulness meditation practice improves electrophysiological markers of attentional control. Front. Hum. Neurosci. **6**, 18 (2012). https://doi.org/10.3389/fnhum.2012.00018

23. Kabat-Zinn, J.: Coming To Our Senses: Healing Ourselves and the World Through Mindfulness. Hyperion, New York (2005)

24. Sliwinski, J., Katsikitis, M., Jones, C.M.: A review of interactive technologies as support tools for the cultivation of mindfulness. Mindfulness (N. Y.) **8**, 1150–1159 (2017). https://doi.org/10.1007/s12671-017-0698-x

25. Lomas, T., Cartwright, T., Edginton, T., Ridge, D.: A qualitative analysis of experiential challenges associated with meditation practice. Mindfulness (N. Y.) **6**, 848–860 (2015). https://doi.org/10.1007/s12671-014-0329-8

26. Cleeremans, A., Jiménez, L.: Implicit Learning and Consciousness: A Graded, Dynamic Perspective. In: French, R.M., Cleeremans, A. (eds.) Implicit Learning and Consciousness: An Empirical, Philosophical and Computational Consensus in the Making, pp. 1–40. Psychology Press, Hove (2002)

27. Dehaene, S., Changeux, J.-P., Naccache, L., Sackur, J., Sergent, C.: Conscious, preconscious, and subliminal processing: a testable taxonomy. Trends Cogn. Sci. **10**, 204–211 (2006). https://doi.org/10.1016/j.tics.2006.03.007

28. Schooler, J.W., Mrazek, M.D., Baird, B., Winkielman, P.: Minding the mind: the value of distinguishing among unconscious, conscious, and metaconscious processes. APA Handbook of Personality and Social Psychology. Attitudes and Social Cognition, vol. 1, pp. 179–202. American Psychological Association, Washington (2015)

29. Dehaene, S.: Consciousness and the Brain: Deciphering How the Brain Codes Our Thoughts. Viking Press, New York (2014)

30. De Tanti, A., Inzaghi, M.G., Bonelli, G., Mancuso, M., Magnani, M., Santucci, N.: Normative data of the MIDA battery for the evaluation of reaction times. Eur. Medicophys. **34**, 211–220 (1998)

31. Folstein, J.R., Van Petten, C.: Influence of cognitive control and mismatch on the N2 component of the ERP: a review. Psychophysiology **45**, 152–170 (2008). https://doi.org/10.1111/j.1469-8986.2007.00602.x

32. Atchley, R., Klee, D., Memmott, T., Goodrich, E., Wahbeh, H., Oken, B.: Event-related potential correlates of mindfulness meditation competence. Neuroscience **320**, 83–92 (2016). https://doi.org/10.1016/j.neuroscience.2016.01.051

33. Tang, Y.-Y., Hölzel, B.K., Posner, M.I.: The neuroscience of mindfulness meditation. Nat. Rev. Neurosci. **16**, 213–225 (2015). https://doi.org/10.1038/nrn3916

34. Crivelli, D. Fronda, G., Venturella, I., Balconi, M.: Supporting mindfulness practices with brain-sensing devices. Cognitive and Electrophysiological Evidences. Mindfulness. https://doi.org/10.1007/s12671-018-0975-3

Chapter 3
Age and Culture Effects on the Ability to Decode Affect Bursts

Anna Esposito, Antonietta M. Esposito, Filomena Scibelli, Mauro N. Maldonato and Carl Vogel

Abstract This paper investigates the ability of adolescents (aged 13–15 years) and young adults (aged 20–26 years) to decode affective bursts culturally situated in a different context (Francophone vs. South Italian). The effects of context show that Italian subjects perform poorly with respect to the Francophone ones revealing a significant native speaker advantage in decoding the selected affective bursts. In addition, adolescents perform better than young adults, particularly in the decoding and intensity ratings of affective bursts of happiness, pain, and pleasure suggesting an effect of age related to language expertise.

Keywords Affective bursts · Age and cultural effects · Universal invariance on vocal emotional expression recognition

A. Esposito (✉)
Dipartimento di Psicologia, Università della Campania "Luigi Vanvitelli", Caserta, Italy
e-mail: iiass.annaesp@tin.it

A. Esposito
IIASS, Vietri Sul Mare, Italy

A. M. Esposito
Istituto Nazionale di Geofisica e Vulcanologia, Sez. di Napoli Osservatorio Vesuviano, Naples, Italy
e-mail: antonietta.esposito@ingv.it

F. Scibelli
Dipartimento di Studi Umanistici, Università di Napoli "Federico II",
Naples, Italy
e-mail: filomena.scibelli@libero.it

M. N. Maldonato
Dipartimento di Neuroscience and Rep. O. Sciences, Università di Napoli "Federico II",
Naples, Italy
e-mail: nelsonmauro.maldonato@unina.it

C. Vogel
School of Computer Science and Statistics, Trinity College Dublin, Dublin, Ireland
e-mail: vogel@tcd.ie

© Springer International Publishing AG, part of Springer Nature 2019
A. Esposito et al. (eds.), *Quantifying and Processing Biomedical and Behavioral Signals*, Smart Innovation, Systems and Technologies 103,
https://doi.org/10.1007/978-3-319-95095-2_3

23

3.1   Introduction

This paper is positioned inside the lively debate on whether emotions are universally shared and correctly decoded among humans versus whether their recognition and perception are strongly shaped by the context (the social, physical, and organizational context [3, 4]). Discussions on this issue have mainly involved only emotional facial expressions ([8] vs. [6]). Vocal emotional expressions and affective bursts have been largely excluded. The more recent emotional vocal data are in favor of a universally shared vocal code slightly affected by language specificities [12], even though experiments exploiting ecological data suggest more language and context specific effects [5, 7, 9, 14]. To assess cultural and age specific effects on the decoding of affective bursts, the reported experiment recruited two groups of Italian participants (adolescents and young adults) and proposed them to assess the Montreal affective bursts collected by Belin et al. [1].

As for this paper's structure, the following section reports on materials and experimental set-up. Section 3.3 describes the experimental results. Discussion and conclusions are provided in Sects. 3.4 and 3.5 respectively.

3.2   Material and Experimental Procedures

3.2.1   Material

The exploited stimuli consist of 90 affect bursts constituting what is known as "the Montreal database of affective bursts".[1] The involved emotions were those of anger, disgust, fear, pain, sadness, surprise, happiness, and pleasure (plus a neutral expression), produced by 10 different Francophone actors (5 male and 5 female) and assessed by 30 Francophone participants both on labeling and intensity accuracy (details are provided in Belin et al. [1]).

3.2.2   Participants

Two differently aged groups of participants (adolescents and young adults) were involved in the experiment. Forty-six adolescents, equally balanced for gender, and aged between 13 and 15 years (mean age = 14.4 years; standard deviation = ±0.5) were recruited in the high school "Francesco Durante" situated in Frattamaggiore, Napoli, Italy. Before starting the data collection, the experiment was approved by both the dean and ethical committee of the school. Subsequently, also the approval

---

[1] See vnl.psy.gla.ac.uk/resources.php (last verified—January 2018). In particular, see: vnl.psy.gla.ac.uk/sounds/ and search for Montreal_Affective_Voices.zip (last verified—January 2018).

of parents was obtained. The parents of the adolescents were first debriefed on the experimental procedure and then required to sign a consent form in order to allow their child to be involved in the data collection. Forty-six young adults, equally balanced by gender, and aged between 20 and 26 years (mean age $= 22$ years; standard deviation $= \pm 1.9$) were recruited at the Università della Campania "Luigi Vanvitelli", located in Caserta, Italy. Students in Psychology were excluded from the experiment. Before starting the data collection, participants were asked to sign a consent form where they expressly declared to voluntarily participate and after the data collection they were debriefed on the aims and goals of the experiment. Both adolescents' parents and young adults were informed that they were free to leave the experimental procedure at any time and their data were anonymized for privacy protection.

3.2.3   Procedures

Each participant was asked to listen the 90 affect bursts contained in the Montreal Affective Voice database. The stimuli were administered randomly through computer headsets. The randomization was made exploiting the "Superlab 4.0" software. Participants were first required to label (with one of the following labels: anger, disgust, fear, pain, sadness, surprise, happiness, pleasure and neutral) the auditory stimulus, and then rate, on a Likert scale from 0 to 9 the perceived intensity of the stimulus. For example, if the listened stimulus was labeled "anger" then the corresponding perceived intensity of the stimulus would have been rated from $0 =$ "not at all angry" to $9 =$ "extremely angry".

3.3   Results

The confusion matrices obtained on the decoding accuracy of the two groups are reported in Tables 3.1 and 3.2 for adolescents and young adults respectively. The reported percentage values are approximated to rounded integer values.

Data were analyzed through a mixed ANOVA analysis (Bonferroni post hocs, and alpha $= .05$) with adolescents, young adults (groups), and gender as between subject variables and emotional affect burst categories as within subject variables. The recognition accuracy was found to be significantly different between adolescents and young adults ($F(1,88) = 7.01$; $p = .010$; partial eta$^2 = .074$) with adolescents performing better. Decoding accuracy was significantly different among affect burst emotional categories ($F(8,704) = 84.08$; $p < .001$; partial eta$^2 = .48$). In particular, happy affect bursts were significantly better decoded than other affect burst emotional categories (Bonferroni post hoc, $p < .001$) except for the neutral one ($p = 1.00$). No interaction was found between groups and affect burst emotional categories ($F(8,720) = 1.44$; $p = .19$; partial eta$^2 = .016$). Significant differences between adolescents and young adults (Bonferroni post hoc) were found for happy ($p = .042$),

Table 3.1 Confusion matrix on the percentage (%) of adolescents' decoding accuracy for the listened affect bursts

Decoding target labels	% of adolescent actual response to target	Decoding response labels									
		Happiness	Fear	Anger	Surprise	Sadness	Disgust	Neutral	Pleasure	Pain	Other emotion
	Happiness	**90**			1			1	4		2
	Fear	1	**49**	10	3	1	3	6	5	17	5
	Anger	1	12	**45**	9		2	3	3	16	8
	Surprise	1	28	2	**27**		6	15	3	12	6
	Sadness	2	1	2		**76**	1	1	1	15	2
	Disgust	1			5	1	**69**	8	7	2	6
	Neutral	1	1		3		2	**84**	2	1	5
	Pleasure		1		13	2	3	7	**63**	2	9
	Pain		5	11	3	3	9	4	9	**52**	4

Table 3.2 Confusion matrix on the percentage (%) of young adults decoding accuracy for the listened affect bursts

Decoding target labels	% of young adults actual response to target	Decoding Response Labels									
		Happiness	Fear	Anger	Surprise	Sadness	Disgust	Neutral	Pleasure	Pain	Other emotion
	Happiness	**78**	1			1	1	2	14	1	2
	Fear	1	**49**	7	6	1	1	14	5	10	7
	Anger		15	**43**	8	1	1	11	2	10	9
	Surprise	2	28	2	**19**		3	24	5	10	7
	Sadness	4	2	1	1	**66**	1	2		22	2
	Disgust	1	2	1	4		**62**	10	7	4	10
	Neutral		1		1	1	1	**86**	2	2	6
	Pleasure	3	2	1	14	2	2	14	**52**	2	8
	Pain	1	10	17	3		7	10	6	**40**	7

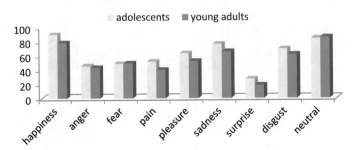

Fig. 3.1 Emotion **correct decoding accuracy** for adolescents and young adults

painful ($p = .020$) and pleasurable ($p = .044$) affect bursts. No significant differences were found between male and female subjects (F(1,88) = .20; $p = .20$; partial eta^2 = .018). A groups*gender interaction showed significant differences between adolescents and young adults males (F(1,88) = 5.00; $p = .28$; partial eta^2 = .054) mostly due to adolescents' ability to better decode affect bursts of pain ($p = .022$). A gender*emotion interaction showed significant differences between male and female only for pleasurable affect bursts (F(1,88) = 4.96; $p = .028$; partial eta^2 = .053). For sake of clarity the percentage of correct decoding accuracy is illustrated in Fig. 3.1 for each affect burst category.

Tables 3.3 and 3.4 display confusion matrices obtained on the two groups' responses for the mean intensity rating percentage of the portrayed emotional affective bursts. Mean rating intensity values were computed, for each emotion, as the ratio of the sum of all the correctly labeled intensity ratings over the sum of both the correctly and incorrectly labeled intensity ratings, multiplied for 100. The reported values are approximated to rounded integer values.

A mixed ANOVA analysis was performed (Bonferroni post hoc, and alpha = .05) on the mean intensity rating percentages, with adolescents, young adults, and gender as between subject variables and affect burst emotional categories as within subject variables. Mean intensity rating percentages were found to be significantly different between adolescents and young adults (F(1,88) = 5.39; $p = .022$; partial eta^2 = .058), with adolescents attributing higher intensity values to the listened affect bursts. Mean intensity rating percentages were also found to be significantly different among affect burst categories (F(8,704) = 69.34; $p < .001$; partial eta^2 = .41). In particular, intensity rating percentages of happy affect bursts were significantly different with respect to other emotional categories (Bonferroni post hoc, $p < .001$), except for the neutral one ($p = 1.00$). No interaction was found between groups and affect burst categories (F(8,704) = 1.84; $p = .08$; partial eta^2 = .020). Adolescents' and young adults mean intensity rating percentages (Bonferroni post hoc) were significantly different for happy ($p = .045$), painful ($p = .018$) and pleasurable ($p = .042$) affect bursts.

No significant differences were found between male and female subjects (F(1,88) = 1.26; $p = .26$; partial eta^2 = .014). A groups*gender interaction showed significant difference between adolescents and young adults males (F(1,88) = 5.33; $p = .23$; partial eta^2 = .057) mostly due to adolescents' ability to attribute higher

Table 3.3 Confusion matrix on the percentage (%) of adolescents' intensity rating values attributed to the listened affect bursts

Intensity target labels	Intensity response labels									
% of adolescent actual response to target	Happiness	Fear	Anger	Surprise	Sadness	Disgust	Neutral	Pleasure	Pain	Other emotion
Happiness	**91**			1			1	4		2
Fear	1	**50**	11	3	1	3	5	5	17	5
Anger	1	12	**46**	8		2	3	3	16	7
Surprise		30	2	**28**		6	12	4	13	5
Sadness	2	1	2		**78**	1		1	14	1
Disgust	1			5	1	**71**	7	8	2	6
Neutral	1	1		4	1	2	**84**	1	1	5
Pleasure	1	1	1	13	1	3	6	**66**	1	8
Pain		5	10	3	3	9	2	10	**53**	4

Table 3.4 Confusion matrix on the percentage (%) of young adult's intensity rating values attributed to the listened affect bursts

Intensity target labels	% of young adults actual response to target	Intensity response labels									
		Happiness	Fear	Anger	Surprise	Sadness	Disgust	Neutral	Pleasure	Pain	Other emotion
	Happiness	**82**	1			1	1	1	11	1	1
	Fear	1	**55**	8	6	1	1	8	5	10	6
	Anger		15	**45**	7	1	1	8	3	11	9
	Surprise	2	30	2	**21**	1	2	18	6	11	7
	Sadness	4	2	1	1	**71**		1	7	17	2
	Disgust	1	1	1	4		**64**	7	7	4	10
	Neutral	1	1		1	1	1	**84**	2	3	6
	Pleasure	3	1	1	13	2	2	10	**58**	2	7
	Pain	1	9	17	3		7	5	7	**44**	6

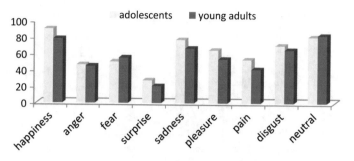

Fig. 3.2 Mean intensity rating percentages for adolescents and young adults

rating percentages to affect bursts of pain ($p = .012$). A gender*emotion interaction showed significant differences between male and female only for pleasurable affect bursts ($F(1,88) = 5.58$; $p = .020$; partial eta$^2 = .060$). For sake of clarity the mean intensity rating percentages are illustrated in Fig. 3.2 for each affect burst category.

3.4   Discussion

Affect bursts are nonverbal emotional vocalizations defined as *"short, emotional nonspeech expressions, comprising both clear nonspeech sounds (e.g., laughter) and interjections with a phonemic structure (e.g., 'Wow!'), but excluding 'verbal' interjections that can occur as a different part of speech (like 'Heaven!,' 'No!,' etc.)"* [13, p. 103]. These non-lexical vocalizations are considered genuine and spontaneous emotional expressions tied to our evolutionary heritages [10]. It has been shown that affect bursts expressing anger, fear, sadness. surprise, disgust, and amusement are decoded above chances across and intra cultures (see Schröder [13] for German, Sauter et al. [10] for British, and Sauter et al. [11] comparing the decoding accuracy of Himba and British speakers). Similar data are reported by Belin et al. [1] on affect bursts produced and rated by Francophone speakers. The natural inference made from these results was that affect bursts are cross-culturally decoded as emotional states. In addition, *"the emotions found to be recognized from* [affect bursts] *correspond to those universally inferred from facial expressions of emotions supporting theories proposing that these emotions are psychological universals and constitute a set of basic, evolved functions that are shared by all humans"* [11, p. 2411]. Our data shows that Italian adolescents and young adults significantly differ in their ability to correctly decode and rate the intensity of non-native affect bursts of happiness, pain, and pleasure suggesting an effect of age even among these close aged groups. Our suggested explanation is that the more is the subject's language experience, the less is her ability to capture emotional information in non-native vocalizations. Although adolescents have less experience with their language, for the results here, this suggests that perhaps they have internalized fewer of the cultural modes of affective burst

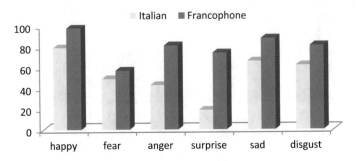

Fig. 3.3 Italian versus Francophone young speakers decoding accuracy

expressions corresponding to their language community. This possibly explains why they are better at decoding them. It further suggests exploring bilinguals in the same age groups given that the cross-cultural contact would be controlled in the bilingual setting. Language expertise is a strong cultural bias that reinforces preferences for familiar and culturally situated affective vocalizations.

Figure 3.3 illustrated the ability of Italian and Francophone young speakers to correctly decode affective bursts associated with so called primary emotions [2]. The Francophone data are those reported in Table 3.4 of the pdf file that can be downloaded from http://vnl.psy.gla.ac.uk/resources.php. It must be underlined that the reported comparisons do not include all the affective bursts under examination since the data supplied by Belin et al. [1] do not provide such information. Therefore, comparisons are reported only for the correct decoding accuracy of happiness, fear, anger, surprise, sadness, and disgust.

Figure 3.3 illustrates how Italian and Francophone subjects significantly differ in their ability to decode affect bursts of primary emotions (one tailed t-test for independent means, $t = 2.61749$, $p = .012$), with Italian participants performing significantly worse than Francophones for fear, anger, and surprise.

Figure 3.4 illustrated the mean intensity rating percentages obtained from Italian and Francophones speakers on the affect bursts under examination. The Francophone data are reported in Belin 2008 (Table 3, p. 535). As it can be observed from Fig. 3.4, Italian and Francophone subjects significantly differ in their ability to rate the intensity level of the listened affective bursts (one tailed t-test for independent means, $t = 2.2452$, $p = .020$), with Italian performing significantly worse for anger, surprise, and pain. There is a clear native speaker advantage both in decoding and rating the intensity level of the proposed bursts. These failures underlie the need of more comparable investigations on how cultures affect the human ability to decode affect bursts. This evidence suggests that both language experiences (adolescents are less experienced than young adults on their native language) and specific cultural differences (those that may differentiate Italian form Francophone subjects belonging to the same Western culture) play a role. The influence of these factors cannot be ignored when "*The goal is to provide experimental and theoretical models of behav-*

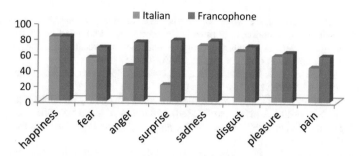

Fig. 3.4 Italian (young adults) versus Francophone speakers mean intensity ratings

iors for developing a computational paradigm that should produce [ICT interfaces] equipped with a human level [of] automaton intelligence" [4, p. 48].

3.5   Conclusion

Emotional vocalizations and vocal expressions of emotions have received less attention in the literature than visual expressions. Perhaps this is because visual cues are thought to have a greater candidacy for cross-linguistic universality. However, the present work contributes evidence that vocal expressions of some emotion categories are also candidates for universal status. Of course, demonstration of substantial agreement between two languages and age groups is far from demonstrating universality. This work establish a *prima facie* case for further explorations.

Acknowledgements The research leading to the results presented in this paper has been conducted in the project EMPATHIC (Grant No.: 769872) that received funding from the European Union's Horizon 2020 research and innovation programme. The dean and the ethical committee of the "Francesco Durante" school situated in Frattamaggiore, Napoli, Italy are acknowledged for allowing the data collection. Acknowledgements are also due to parents, adolescents, and young adults participating to the experiment.

References

1. Belin, P., Fillion-Bilodeau, S., Gosselin, F.: The Montreal Affective Voices: a validated set of nonverbal affect bursts for research on auditory affective processing. Behav. Res. Methods **40**(2), 531–539 (2008)
2. Ekman, P.: Emotions Revealed: Recognizing Faces and Feelings to Improve Communication and Emotional Life. Weidenfeld and Nicolson, London (2003)
3. Esposito, A., Jain, L.C.: Modeling social signals and contexts in robotic socially believable behaving systems. In: Esposito, A., Jain, L.C. (eds.) Toward Robotic Socially Believable Behaving Systems Volume II—"Modeling Social Signals". ISRL Series, vol. 106, pp. 5–13. Springer International Publishing Switzerland (2016)

4. Esposito, A., Esposito, A.M., Vogel, C.: Needs and challenges in human computer interaction for processing social emotional information. Pattern Recogn. Lett. **66**, 41–51 (2015)
5. Esposito, A., Esposito, A.M.: On the recognition of emotional vocal expressions: motivations for an holistic approach. Cogn. Process. J. **13**(2), 541–550 (2012)
6. Jack, R.E., Schyns, P.G.: The human face as a dynamic tool for social communication. Curr. Biol. **25**(14), R621–R634 (2015)
7. Maldonato, N.M., Dell'Orco, S.: Making decision under uncertainty, emotions, risk and biases. In: Bassis, S., Esposito, A., Morabito, F.C. (eds.) Advances in Neural Networks: Computational and Theoretical Issues. SIST Series, vol. 37, pp. 293–302. Springer International Publishing Switzerland (2015)
8. Matsumoto, D., Nezlek, J.B., Koopmann, B.: Evidence for universality in phenomenological emotion response system coherence. Emotion **7**(1), 57–67 (2007)
9. Riviello, M.T., Esposito, A.: On the Perception of Dynamic Emotional Expressions: A Cross-Cultural Comparison. In: Hussain, A. (ed.) SpringerBriefs in Cognitive Computation, vol. 6, pp. 1–45 (2016)
10. Sauter, D., Eisner, F., Ekman, P., Scott, S.K.: Perceptual cues in non-verbal vocal expressions of emotion. Q. J. Exp. Psychol. (Hove) **63**(11), 2251–2272 (2010)
11. Sauter, D., Eisner, F., Ekman, P., Scott, S.K.: Cross-cultural recognition of basic emotions through nonverbal emotional vocalizations. PNAS **107**(6), 2408–2412 (2010)
12. Scherer, K.R., Banse, R., Wallbott, H.C.: Emotion inferences from vocal expression correlate across languages and cultures. J. Cross Cult. Psychol. **32**(1), 76–92 (2007)
13. Schröder, M.: Experimental study of affect bursts. Speech Commun. **40**, 99–116 (2003)
14. Troncone, A., Palumbo, D., Esposito, A.: Mood effects on the decoding of emotional voices. In: Bassis, S., et al. (eds.) Recent Advances of Neural Network Models and Applications. SIST, vol. 26, pp. 325–332. International Publishing Switzerland (2014)

Chapter 4
Adults' Implicit Reactions to Typical and Atypical Infant Cues

Vincenzo Paolo Senese, Francesca Santamaria, Ida Sergi
and Gianluca Esposito

Abstract This study investigates the valence of adults' implicit associations to typical and atypical infant cues, and the consistency of responses across the different stimuli. 48 non-parent adults (25 females, 23 males) were presented three kinds of infant cues, typical cry (TD-cry), atypical cry (ASD-cry) and infant faces, and their implicit associations were measured by means of the Single Category Implicit Association Test (SC-IAT). Results showed that, independently of gender, the implicit associations to typical and atypical infant cries had the same negative valence, whereas infant faces were implicitly associated to the positive dimension. Moreover, data showed that implicit responses to the different infant cues were not associated. These results suggest that more controlled processes influence the perceptions of atypical infant cry, and confirm the need to investigate individual reactions to infant cues by adopting a multilevel approach.

Keywords Infant cry · Infant face · ASD · Implicit association · SC-IAT

V. P. Senese (✉) · F. Santamaria · I. Sergi
Department of Psychology, University of Campania "Luigi Vanvitelli", Caserta, Italy
e-mail: vincenzopaolo.senese@unicampania.it

F. Santamaria
e-mail: francescasantamaria44@gmail.com

I. Sergi
e-mail: ida.sergi@unicampania.it

G. Esposito
Division of Psychology, Nanyang Technological University, Singapore, Singapore
e-mail: gianluca.esposito@unitn.it

G. Esposito
Department of Psychology and Cognitive Sciences, University of Trento, Trento, Italy

© Springer International Publishing AG, part of Springer Nature 2019
A. Esposito et al. (eds.), *Quantifying and Processing Biomedical and Behavioral Signals*, Smart Innovation, Systems and Technologies 103,
https://doi.org/10.1007/978-3-319-95095-2_4

4.1 Introduction

Several researchers argue that humans have an innate predisposition to caregiving infants [1] which is the expression of a biologically rooted behaviour [2–4]. Providing a different perspective, studies also showed that adults do not always manifest adequate or sensitive behaviours toward infants, and that child abuse and neglect were also observed [5–7]. Therefore, individual differences can be considered a critical factor in the regulation of infant caregiving [3, 8, 9]. Indeed, even when it is stated that caregiving is influenced by the interaction of a variety of forces (adult characteristics, child characteristics and context characteristics), as in the Determinants of Parenting Model [10], it is recognized that the individuals have a crucial role in determining caregiving behaviours, and that individual characteristics can even buffer the negative effects of the other factors.

In line with these considerations, in the literature, researchers have investigated adults' reaction to salient infant cues in order to better understand the processes that regulate adult-infant interaction, showing that caregiving behaviours relies on the interrelation of multilevel systems that involve cortical and subcortical brain structures [4, 8]; and that responsiveness to infant cues is related to infant later development [11–14].

Among the infant cues, the most studied are infant faces and cries. Infant faces are characterized by a constellation of morphological features, "*Kindchenschema*", which distinguish them from adult faces [15], capture attention [16], are associated with a positive implicit evaluation [3], activate specific brain areas [4, 8], and elicit willingness to approach, to smile and to communicate [17]. These results were consistent across genders [3].

Infant cry is the earliest mean of infant social vocal communication which promotes caregiver proximity, activates specific brain areas [4, 8], and is supposed to trigger caregiving behaviours [18, 19]. Typical infant cry (TD-cry) has a specific acoustic pattern (e.g., pause length, number of utterances, and fundamental frequency) that influences the perception of infant distress and its meaning [20, 21]. Studies investigating whether gender influences adults' responses to infant cries showed a mixed pattern of results [22].

Infant cry also has the advantage of allowing the investigation of adults' responses to typical and atypical infant cues, thus facilitating the examination of the effects of different child characteristics on the individual reactions. Indeed, it has been shown that cry of infants later diagnosed with ASD (ASD-cry) has a specific acoustical pattern (shorter pauses, fewer utterances, and higher fundamental frequency), activates specific brain areas [23], is recognized as different by caregivers [18], and affects adult behaviours [24].

Despite its specificity, only few studies have investigated adults' reactions to atypical infant cry (see [9, 18, 23, 24]). Moreover, the extant studies have the limitation of not evaluating in a direct way the valence of the reaction to this atypical infant cue or taking into account the social desirability bias. Indeed, they mainly focused on self-report or behavioural responses in women, thus taking into account only conscious or controlled processes, or considering indirect measures (e.g., physiological) that cannot clarify the positive or negative valence of the responses. Only one study considered males [9], and no study directly investigated gender differences.

Recently, the Single Category Implicit Association Test (SC-IAT) has been adapted to investigate in a direct way the valence of implicit reaction to visual and acoustical infant cues by taking into account the social desirability bias [3, 22]. Senese and colleagues showed that, independently of gender and parental status, infant faces were associated with specific and positive implicit reactions, whereas TD-cries were associated to negative implicit reaction. Besides, results showed wide individual differences in implicit reactions that in turn were associated to parental models. To our knowledge no study has investigated yet the valence of implicit reaction to ASD-cry, or investigated the association between implicit reactions to different infant cues. From a theoretical perspective, showing if the ASD-cry has a specific implicit valence could be useful to better understand the processes that regulate the different perception of atypical cries [18] and the consequent behaviours [24].

Building on the aforementioned considerations, the aim of the present study was to investigate the valence of the implicit reaction to ASD-cry and to compare adults' reactions to typical and atypical infant cues. To this aim three SC-IATs were adapted to evaluate implicit associations to TD-cry, ASD-cry and typical infant faces. According to the literature [3, 22], we expected negative implicit association to typical and atypical infant cry, positive implicit reaction to infant faces, and significant differences between TD- and ASD-cry [18], with the latter showing a more negative implicit association. In line with the literature on implicit reaction to infant cues [3, 22], no gender differences were expected.

4.2 Method

4.2.1 Sample

A total of 48 non-parent adults (25 females, 23 males) participated in a within-subject experimental design. Their ages ranged from 19 to 38 years ($M=24.94$, $SD=3.5$), and their educational level varied from middle school to college levels. Males and females were matched as a function of the age, $F<1$, and all participants were tested individually.

4.2.2 Procedure

The experimental session was divided in two phases. In the first phase, basic socio demographic information (i.e., sex, age and socio economic status) was collected, after which the three SC-IATs (TD-cry, ASD-cry and infant faces) were administered in a counterbalanced order. The study was conducted in accordance with ethical principles stated in the Helsinki Declaration. All participants signed a written informed consent before starting the experimental session. The session lasted about 25 min.

4.2.3 Measures

Single Category Implicit Association Test (SC-IAT). Abiding by the literature [3, 22], three SC-IATs—two auditory versions and one classical visual version—were adapted to evaluate the valence of implicit reactions to typical and atypical infant cues: TD-cry, ASD-cry, and infant faces. The SC-IAT is a two-stage classification task. In each stage, a single target item (audio clip or picture) was presented with target words in random order. Participants were presented one item at time (i.e., target items or words) and asked to classify it into the correct category as quickly as possible. Words were distinguished as "good" and "bad" and had to be classified into the positive or negative category, respectively. In case of error, an "X" appeared at the centre of the screen. To emphasize speed of responding, a response window of 1500 ms following stimulus onset was applied for each stimulus. Each SC-IAT was repeated twice. In the first stage, good words and target objects were categorized according to the same response key, and bad words were categorized using a second key (positive condition). In the second stage, bad words and target objects were categorized using the same response key, and good words were categorized using the second key (negative condition). The SC-IAT score is derived from the comparison of latencies of responses in the two classification stages. If participants were faster in categorizing stimuli in the positive condition in comparison to the negative condition, they were considered to have positive implicit attitudes towards the target. If the contrary was true, a negative implicit attitude was attributed. For each test, the SC-IAT score was calculated by dividing the difference between means of RTs of the two classification conditions by the standard deviation of latencies of the two phases [25]. Scores around 0 indicate no IAT effect; absolute values from 0.2 to 0.3 indicate a "slight" effect, values around 0.5 a "medium" effect, and values of about 0.8 to infinity a "large" effect. In this study, positive values indicate that the target cue was implicitly associated with the positive dimension. All the SC-IAT scores showed adequate reliability ($\alpha s > .70$).

Stimuli. Infant stimuli were the same used into previous researches. In particular, TD-cry [22] and ASD-cry [18] were extracted from home videos of 13-month-old infants to be acoustically representative of the relative category. Both stimuli lasted 5 s each and were normalized for intensity. They were presented through headphones at a constant volume. Infant faces [3] portrayed infants with a mean age of about 6 months showing a neutral expression.

4.3 Data Analysis

Normality of univariate distributions of SC-IAT scores of each type of cue was preliminarily checked. To analyse the effect of the cue characteristics on the valence of infant cues, implicit associations were analysed by means of a 3×2 mixed *ANOVA* that treated infant Cue (TD-cry, ASD-cry and infant faces) as a three-level within-subjects factor, and Gender (males vs. females) as a two-level between-subjects factor. Bonferroni correction was used to analyse post hoc effects of significant factors, and partial eta squared (η_p^2) was used to evaluate the magnitude of significant effects. To investigate the association of implicit reactions to the different implicit cues, Pearson correlation coefficients were computed.

4.4 Results

The *ANOVA* on SC-IAT scores showed that implicit associations were influenced by the Cue, $F(2,90)=17.56$, $p< .001$, $\eta_p^2 = .281$, not significant were the Gender main effect, $F(1,45)=1.07$, $p= .306$, $\eta_p^2 = .023$, and the Cue×Gender interaction, $F<1$. Post hoc analysis revealed that infant faces, $M=0.194$, 95% CI [0.096; 0.293], were associated with the positive dimension, while the TD-cry, $M=-0.183$, 95%CI [−0.297; −0.068], and the ASD-cry, $M=-0.154$, 95%CI [−0.256; −0.051], were associated with the negative dimension (see Fig. 4.1). No differences were observed between typical and atypical cry.

Finally, correlation analysis showed that the implicit reactions to infant cues were not significantly associated (see Table 4.1).

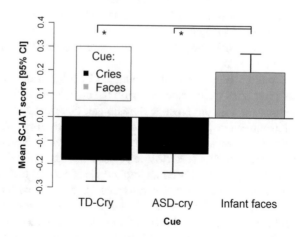

Fig. 4.1 Mean SC-IAT score as a function of the Cue (*$p < .001$)

Table 4.1 Pearson correlation coefficients between implicit associations to infant cues

Variables	1	2	3
SC-IAT			
1. TD-cry	–		
2. ASD-cry	.210	–	
3. Infant faces	−.030	.099	–

4.5 Discussion

The aim of the present study was to assess whether typical and atypical infant cries were associated with specific implicit responses in non-parent adults, as well as the consistency of implicit responses toward different infant cues. The literature showed that the cry of infants diagnosed with ASD has a specific acoustic pattern that activates specific brain areas, is differentiated by caregivers, and affects adults' behaviours [9, 18, 23, 24]. We hypothesized that ASD-cry would be associated with a more negative implicit response if compared with TD-cry and infant faces, with the latter showing a positive implicit association. Moreover, given that previous studies did not show gender differences in the valence of the implicit responses to infant faces and TD-cry [3, 22], we expected that males and females would show a similar implicit association to infant cues. To test these hypotheses, participants were presented three kinds of infant cues (TD-cry, ASD-cry and infant faces) and their implicit associations were measures by means of the SC-IAT paradigm.

We considered infant cry because it has been showed that adult responses to this cue are associated to the quality of caregiver-infant relationships and child development [11–14], and because it serves as a good basis for the investigation of child characteristics on adult responsiveness. We used the SC-IAT paradigm because to our knowledge it is the only paradigm that allows the direct investigation of the valence of responses to infant cues by taking into account the social desirability bias [3, 22]. This is the first study that implements this paradigm to investigate adults' implicit reactions to atypical cries.

In line with the previous literature [18, 22], results showed that both typical and atypical cries were associated with the negative domain, whereas infant faces were associated with the positive domain [3], therefore showing that adults have a specific implicit response to infant faces and negative implicit associations toward infant cries. Contrary to our expectations and the previous literature [18], no significant differences were observed between typical and atypical cries. In their study, Bornstein and colleagues [18] used an explicit classification task showing that women were slower in classifying ASD-cry versus TD-cry. A possible explanation could be that at implicit level the valence of the infant cry is independent of the acoustic pattern, and that the previous observed differences are related to more controlled processes. Indeed, according to parental models [4, 8], adult responses to infant cues are regulated at different levels, from more reflexive processes to more controlled ones. Another possible explanation could be that the way we implemented the SC-IAT paradigm was not sensitive to evaluate the differences between typical and atypical cry. This is the first study that used SC-IAT to evaluate implicit responses to similar auditory stimuli. Future studies should replicate this study by modifying the paradigm to elucidate whether TD-cry and ASD-cry have specific implicit associations.

With regards the gender differences, our results replicate previous findings showing similar implicit responses between males and females on infant faces and TD-cry [3, 22]. Moreover, data showed that no gender differences were observed on ASD-cry. Therefore, this is the first study that directly investigated gender differences in response to atypical infant cry. In the literature studies investigating gender differences on responses to typical infant cues showed a mixed pattern (see [3, 22]). It is possible that gender differences are the expression of conscious or controlled thoughts and beliefs. Further studies that investigate adults' responses to infant cues by using a multilevel approach are needed.

Finally, the associations between implicit responses to the different infant cues were investigated. Results showed a substantial independence between responses to cry (both typical and atypical) and infant faces. In the literature only one study [26] directly compared reaction (P300) to different infant cues and showed differential responses as a function of the infant cue. Further studies are needed to investigate to what extent responses to different infant cues are the expression of a general caregiving propensity or reflect different components.

The results of this study should be interpreted with certain limitations in mind. First, we considered only a small sample ($N=48$), therefore it is possible that the results neglected small effects size. Further studies with bigger samples should replicate the investigation and test the robustness of our findings. Second, we considered only non-parent adults because we were interested in the investigation of adults' responsiveness to infant cues independent of parental experience. Further studies should directly compare parents and non parents on implicit responses to typical and atypical infant cues. Third, we administered infant cues by adopting a unimodal methodology (acoustic vs. visual), while research showed that the multimodal approach is a more valid methodology for investigating individual responses because human perception is holistic [27]. Further studies should replicate the findings by adopting a more ecological multimodal approach that integrates at least audio and visual information. Fourth, we measured implicit reactions only, but researchers agree that caregiving behaviours are regulated at different levels of processing. Further studies applying a multilevel approach should be carried out. Finally, we considered the valence of implicit reactions to different infant cues, but no direct measure of caregiving was considered. Further study should include a direct measure of caregiving to investigate the predictive validity of the implicit reactions.

4.6 Conclusions

The result of this study showed that, independent of gender, implicit responses to typical and atypical infant cries have the same negative valence, while it confirmed that infant faces have a positive valence. Moreover, the data showed wide individual differences and that implicit responses to the different infant cues were not associated. If we assume that the valence of adults' implicit associations to infant cues may contribute to influencing the quality of adult-infant interaction, and consequent child development, then we may suggest that the evaluation of adult implicit associations to different infant cues should be included in the screening protocols in order to better prevent negative outcomes and to plan well-tailored intervention programs aimed at facilitating the expression of sensitive caregiving towards infants.

References

1. Papoŭsek, H., Papoŭsek, M.: Intuitive parenting. In: Bornstein, M.H. (ed.) Handbook of Parenting, 2nd edn, vol. 2, pp. 183–203. Lawrence Erlbaum Associates, Mahwah, NJ (2002)
2. Bornstein, M.H.: Determinants of Parenting. Dev. Psychopathol. Four **5**, 1–91 (2016)
3. Senese, V.P., De Falco, S., Bornstein, M.H., Caria, A., Buffolino, S., Venuti, P.: Human infant faces provoke implicit positive affective responses in parents and non-parents alike. PlosOne **8**(11), e80379 (2013). https://doi.org/10.1371/journal.pone.0080379
4. Swain, J.E., Kim, P., Spice, J., Ho, S.S., Dayton, C.J., Elmadih, A., Abel, K.M.: Approaching the biology of human parental attachment: brain imaging, oxytocin and coordinated assessments of mothers and fathers. Brain Res. **1580**, 78–101 (2014)
5. Barnow, S., Lucht, M., Freyberger, H.-J.: Correlates of aggressive and delinquent conduct problems in adolescence. Aggressive Behav. **31**, 24–39 (2005)
6. Beck, J.E., Shaw, D.S.: The influence of perinatal complications and environmental adversity on boys' antisocial behavior. J. Child Psychol. Psychiatry **46**, 35–46 (2005)
7. Putnick, D.L., Bornstein, M.H., Lansford, J.E., Chang, L., Deater-Deckard, K., Di Giunta, L., Bombi, A.S.: Agreement in mother and father acceptance-rejection, warmth, and hostility/rejection/neglect of children across nine countries. Cross Cult. Res. **46**, 191–223 (2012). https://doi.org/10.1177/1069397112440931
8. Barrett, J., Fleming, A.: Annual research review: all mothers are not created equal: neural and psychobiological perspectives on mothering and the importance of individual differences. J. Child Psychol. Psychiatry **52**(4), 368–397 (2011)
9. Esposito, G., Valenzi, S., Islam, T., Bornstein, M.H.: Three physiological responses in fathers and non-fathers' to vocalizations of typically developing infants and infants with Autism Spectrum Disorder. Res. Dev. Disabil. **43–44**, 43–50 (2015). https://doi.org/10.1016/j.ridd.2015.06.007
10. Belsky, J.: The determinants of parenting: a process model. Child Dev. **55**, 83–96 (1984)
11. Higley, E., Dozier, M.: Night time maternal responsiveness and infant attachment at one year. Attachment Hum. Dev. **11**(4), 347–363 (2009)
12. Kim, P., Feldman, R., Mayes, L.C., Eicher, V., Thompson, N., Leckman, J.F., Swain, J.E.: Breastfeeding, brain activation to own infant cry, and maternal sensitivity. J. Child Psychol. Psychiatry **52**, 907–915 (2011)
13. Leerkes, E.M., Parade, S.H., Gudmundson, J.A.: Mothers' emotional reactions to crying pose risk for subsequent attachment insecurity. J. Fam. Psychol. **25**(5), 635–643 (2011). https://doi.org/10.1037/a0023654
14. McElwain, N.L., Booth-Laforce, C.: Maternal sensitivity to infant distress and nondistress as predictors of infant-mother attachment security. J. Fam. Psychol. **20**(2), 247–255 (2006)
15. Lorenz, K.Z.: Studies in Animal and Human Behaviour, vol. 2. Methuen & Co., London (1971)
16. Brosch, T., Sander, D., Scherer, K.R.: That baby caught my eye… Attention capture by infant faces. Emotion **7**(3), 685–689 (2007)
17. Caria, A., de Falco, S., Venuti, P., Lee, S., Esposito, G., et al.: Species-specific response to human infant faces in the premotor cortex. NeuroImage **60**(2), 884–893 (2012)

18. Bornstein, M.H., Costlow, K., Truzzi, A., Esposito, G.: Categorizing the cries of infants with ASD versus typically developing infants: a study of adult accuracy and reaction time. Res. Autism Spectr. Disord. **31**, 66–72 (2016)
19. Zeifman, D.M.: An ethological analysis of human infant crying: answering Tinbergen's four questions. Dev. Psychobiol. **39**(4), 265–285 (2001)
20. Esposito, G., Venuti, P.: Developmental changes in the fundamental frequency (f0) of infants' cries: a study of children with Autism Spectrum Disorder. Early Child Development and Care **180**(8), 1093–1102 (2010)
21. Esposito, G., Nakazawa, J., Venuti, P., Bornstein, M.H.: Componential deconstruction of infant distress vocalizations via tree-based models: a study of cry in autism spectrum disorder and typical development. Res. Dev. Disabil. **34**(9), 2717–2724 (2013). https://doi.org/10.1016/j.ri dd.2013.05.036
22. Senese, V.P., Venuti, P., Giordano, F., Napolitano, M., Esposito, G., Bornstein, M.H.: Adults' implicit associations to infant positive and negative acoustic cues: moderation by empathy and gender. Q. J. Exp. Psychol. **70**(9), 1935–1942 (2017). https://doi.org/10.1080/17470218.2016. 1215480
23. Venuti, P., Caria, A., Esposito, G., De Pisapia, N., Bornstein, M.H., de Falco, S.: Differential brain responses to cries of infants with autistic disorder and typical development: an fMRI study. Res. Dev. Disabil. **33**(6), 2255–2264 (2012). https://doi.org/10.1016/j.ridd.2012.06.011
24. Esposito, G., Venuti, P.: Comparative analysis of crying in children with autism, developmental delays, and typical development. Focus Autism Other Dev. Disabil. **24**(4), 240–247 (2009)
25. Greenwald, A.G., Nosek, B.A., Banaji, M.R.: Understanding and using the implicit association test: I. An improved scoring algorithm. J. Pers. Soc. Psychol. **85**, 197–216 (2003)
26. Rutherford, H.J.V., Graber, K.M., Mayes, L.C.: Depression symptomatology and the neural correlates of infant face and cry perception during pregnancy. Soc. Neurosci. **11**(4), 467–474 (2016). https://doi.org/10.1080/17470919.2015.1108224
27. Iachini, T., Maffei, L., Ruotolo, F., Senese, V.P., Ruggiero, G., Masullo, M., Alekseeva, N.: Multisensory assessment of acoustic comfort aboard metros: an Immersive Virtual Reality study. Appl. Cogn. Psychol. **26**, 757–767 (2012). https://doi.org/10.1002/acp.2856

Chapter 5
Adults' Reactions to Infant Cry and Laugh: A Multilevel Study

Vincenzo Paolo Senese, Federico Cioffi, Raffaella Perrella and Augusto Gnisci

Abstract Starting from the assumption that caregiving behaviours are regulated at different levels, the aim of the present paper was to investigate adults' reaction to salient infant cues by means of a multilevel approach. To this aim, psychophysiological responses (Heart Rate Variability), implicit associations (SC-IAT-A), and explicit attitudes (semantic differential) toward salient infant cues were measured on a sample of 25 non-parents adults (14 females, 11 males). Moreover, the trait anxiety and the individual noise sensitivity were considered as controlling factors. Results showed that adults' responses were moderated by the specific measure considered, and that responses at the different levels were only partially consistent. Theoretical and practical implications were discussed.

Keywords Infant cues · Parenting · Heart rate variability · Implicit association SC-IAT

5.1 Introduction

Human infants are characterized by a prolonged dependence from caregivers because they are not self-sufficient, and their survival depends on adequate caregiving [1]. For this reason a series of signals (e.g., cries, laughs) have evolved to help the infant communicate with the environment since birth [1, 2].

V. P. Senese (✉) · F. Cioffi · R. Perrella · A. Gnisci
Department of Psychology, University of Campania "Luigi Vanvitelli", Caserta, Italy
e-mail: vincenzopaolo.senese@unicampania.it

F. Cioffi
e-mail: federico.cioffi.88@gmail.com

R. Perrella
e-mail: raffaella.perrella@unicampania.it

A. Gnisci
e-mail: augusto.gnisci@unicampania.it

© Springer International Publishing AG, part of Springer Nature 2019
A. Esposito et al. (eds.), *Quantifying and Processing Biomedical and Behavioral Signals*, Smart Innovation, Systems and Technologies 103,
https://doi.org/10.1007/978-3-319-95095-2_5

Among the infant cues, the cry is particularly salient because it can act as a trigger for caregiving in adults [2]. However, studies showed that individuals manifest different reactions to this cue, and in some cases infant cry can be associated to maltreatment [3, 4]; indicating that infant cues per se are not enough to guarantee an adequate or sensitive caregiving, but the environmental responses depend on the adult's tendency to promptly recognize the cues and respond appropriately [1, 5, 6]. This is particularly critical because the quality of adults' responsiveness is related to infant later development [7–9].

Given its relevance, recent researches have focused on the investigation of the processes and the factors that regulate adult responsiveness to infant cry by adopting different methodologies [1, 5, 10]. To sum main results, Swain and colleagues [10] have proposed the Parental Brain Model (PBM). According to PBM, once perceived by the sensory cortex, the infant cues interact with three cortico-limbic modules that regulate caregiving behaviour. The three structures represent different levels of processing (reflexive, cognitive, and emotional), and the final behaviour is the result of their interaction. Therefore, the caregiving behaviour is regulated at different levels, from more reflexive processes to more controlled ones.

In the literature, researches using fMRI showed that infant cry activates brain areas involved in parenting [10], and that the activity of amygdala and frontal cortex in response to infant cry was correlated with maternal sensitivity [7]. Similar responses were observed for males and females, but studies showed also gender differences in brain activity associated with infant cry [11, 12].

Studies focusing on Heart Rate Variability (HRV), highlighted that HRV is associated to the quality of caregiving [13–15], though the literature is contradictory; showing that greater HRV is associated to both child abuse and adequate caregiving [15]. As regards gender differences, similar divergent patterns were observed. From one side, no gender differences were observed (e.g., [16, 17]); on the other, researches showed greater HRV heart rate responses to infant cry in males, rather than females (e.g., [18, 19]).

Studies considering explicit self-reported evaluations of infant cry showed that adults reported distress toward infant cry [20], can differentiate the features of the infant cry, and that males tend to report higher distress than females [21].

Finally, as regards the emotional processing of infant cry, to our knowledge there is only one study that directly investigated the valence of the implicit associations [22]. In their study, authors adapted the Single Category Implicit Association Test to evaluate the valence of implicit responses to infant cries and laughs; because previous studies showed that implicit reactions to visual infant cues were associated with parental models [23]. Results showed a weak negative reaction to infant cry, and a positive implicit association to infant laugh. No gender differences were observed [22, 23].

In summary, studies that investigated adults' responses to infant cry showed divergent results. Moreover, they present the limit of not considering at the same time the three levels of processing hypothesized in the PBM and their associations.

Given these considerations, the aim of this study was to investigate adults' reactions to infant cues by taking into account different levels of processing, and to investigate the consistency of responses. In particular, infant cry and laugh were considered as stimuli, and the HRV, the valence of the implicit associations, and explicit evaluations were measured. Because HRV is strongly influenced by anxiety [24], the trait anxiety was controlled in the analysis on psychophysiological responses. Moreover, because several studies showed that noise sensitivity influences sound perception and evaluations, and that males and females showed robust differences on this dimension [22, 25], gender effects were controlled for individual noise sensitivity. We expected gender differences, with females showing more positive indices, on physiological and explicit responses, and a significant association between the different responses. Moreover, we also expected that the gender differences were attenuated or disappeared when the noise sensitivity was considered.

5.2 Methods

5.2.1 Sample

A total of 25 non-parents adults (14 females, 11 males) participated in a within-subject experimental design. Participants age ranged from 19 to 33 years (M age $=$ 23.2 years, $SD = 3.1$). Males and females were matched as a function of the age, $F < 1$, and the socio economic status, $F(1,23) = 1.9, p = .184$.

5.2.2 Procedure

Participants signed a written informed consent before starting the experimental session. The experimental session lasted about 45 min and was divided into three phases. In the first phase, the basic socio-demographic information and the psychophysiological reaction to infant cues were recorded. In the second phase, the valence of the implicit reactions to infant cues was measured. Whereas, in the last phase, participants were administered three self-report scales: a semantic differential scale, the noise sensitivity scale and the trait anxiety inventory. The study was conducted in accordance with the Helsinki declaration.

5.2.3 Measures

5.2.3.1 Psychophysiological Activity Measures

Infant cues paradigm. The Infant cues were administered in a block design. The session was divided into five blocks. In the first block (3 min), participants were seated in front of a computer screen, the electrodes were attached, and they were presented visual instructions that asked to stay relaxed. In the second block (20 s), no stimuli were presented to get a reference measure. In the third block (33 s), the 6 infant cues of the first category were presented in a random order. In the fourth block (20 s), no stimuli were presented. Finally, in the last block (33 s), the 6 infant cues of the second category were presented in a random order. The ECG was continuously recorded in the whole session, and no action was asked to participants to avoid registration artifacts. The order of infant cues blocks was randomized across participants.

Infant cues. Infant cues were the same used into a previous research [22]. They were extracted from home videos of 13-month-old infant to be acoustically representative of typical infant cry and laugh. Stimuli lasted 5 s each, and were normalized for intensity. Cries and laughs were matched as a function of the fundamental frequencies, and the peak amplitudes. They were presented through headphones at a constant volume.

Heart Rate Variability (HRV). The ECG signal was measured with Procomp Infiniti monitoring system, using three disposable pre-gelled electrodes placed on participants' chest. Collected data were then processed using the software ARTiiFACT [26] to remove artifacts and extract, for each block, measures of heart rate variability. In particular, the median inter-beat-interval (IBI) was considered. Moreover, to get an index of variability that was independently of individual differences, the IBI of each test block was subtracted to the IBI of the baseline, so that higher scores indicated an acceleration of heart rate in respect to the baseline.

5.2.3.2   Implicit Measures

Single Category Implicit Association Test. Following the literature [22], two auditory versions of the Single Category Implicit Association Test (SC-IAT-A) were adapted to evaluate the valence of adults' implicit reaction to the infant cues: cries and laughs. In each SC-IAT-A, target stimuli of a single category (cries or laughs) and positive and negative words were presented auditorily in a random order. Participants were asked to classify each item in the respective category as fast and accurate possible by pressing the key associated with the category. In a first block the target items and the positive words were classified with the same key, whereas the negative words with a different key (positive condition). In a second block the target items and the negative words were classified with the same key, whereas the positive words with a different key (negative condition). The SC-IAT scores were calculated by dividing the differences in the latency of responses between the positive and negative conditions by the standard deviation of the latencies in the two conditions [27]. If the subjects were faster in categorizing the stimuli in the positive condition, they were supposed to have a positive implicit attitude toward the target stimuli. Both SC-IAT-A showed adequate reliability ($\alpha > .70$).

5.2.3.3 Self-report Measures

Semantic Differential (SD). To measure the explicit attitude toward babies, a semantic differential scales was administered [28]. In the semantic differential scale, participants were asked to evaluate the "baby" by using ten bipolar adjectives on a seven-point scale. A composite total score was computed, with greater values indicating a positive attitude toward babies. The scale showed adequate reliability ($\alpha = .82$).
The State-Trait Anxiety Inventory (STAI). To get a measure of stable individual anxiety, the trait form of the STAI [29] was administered. The scale presents 20 items describing anxiety-related symptoms and participants were asked to indicate how each item reflects their feelings on a 4-point Likert scale. A composite total score was computed, with higher scores indicating greater anxiety. The scale showed adequate reliability ($\alpha = .78$).
Weinstein Noise Sensitivity Scale (WNSS). In order to assess noise sensitivity, the WNSS [25] was administered. The WNSS is a 20-item self-report scale that evaluates individual's attitude toward typical environmental sounds. For each item, participants were asked to evaluate their agreement on a 6-points Likert scale. A composite total score was computed, with higher scores indicating higher noise sensitivity. The scale showed adequate reliability ($\alpha = .89$).

5.2.4 Data Analysis

To investigate the effect of the type of cue and the gender on psychophysiological activity (HRV), a 2×2 mixed *ANCOVA* that treated infant Cue (cry vs. laugh) as a two-level within-subjects factor, Gender (males vs. females) as a two-level between-subjects factor, and STAI scores as covariate was executed. Moreover, to investigate whether gender differences were influenced by individual noise sensitivity, the analysis was replicated by adding WNSS scores as covariate.

To investigate the effect of the type of cue and the gender on implicit scores (SC-IAT-As), a 2×2 mixed *ANOVA* that treated infant Cue (cry vs. laugh) as a two-level within-subjects factor, and Gender (males vs. females) as a two-level between-subjects factor was executed. The analysis was replicated by adding WNSS scores as covariate.

To investigate the effect of the gender on explicit evaluations (SD), a 2×2 mixed *ANOVA* that treated infant Cue (cry vs. laugh) as a two-level within-subjects factor, and Gender (males vs. females) as a two-level between-subjects factor was executed. The analysis was replicated by adding WNSS scores as covariate. In all analysis Bonferroni correction was used to analyse post hoc effects of significant factors, and partial eta squared (η_p^2) was used to evaluate the magnitude of significant effects.

Finally, to investigate the consistency of responses to infant cues across the different levels of processing, Pearson correlation coefficients between measures were computed. Moreover, partial correlations were also computed to control the associations by gender and anxiety.

5.3   Results

5.3.1   Psychophysiological Responses

The *ANCOVA* conducted on the HRV showed that psychophysiological responses were influenced by the Cue, $F(1,22)=5.80$, $p=.025$, $\eta_p^2=.209$, and the Cue× Gender interaction, $F(1,22)=4.83$, $p=.039$, $\eta^2 p=.180$. Not significant were the Gender, $F(1,22)=1.30$, $p=.266$, $\eta_p^2=.056$, and the STAI main effects, $F<1$. Independently of anxiety, both stimuli were related to a heart rate deceleration in respect to the baseline; but infant laughs were associated with a higher deceleration, $M=-7.26$, 95% CI [−19.42; 4.89], than infant cries, $M=-3.41$, 95% CI [−18.58; 11.76]. The post hoc analysis of the Cue × Gender interaction showed that, independently of anxiety, gender differences on physiological responses were observed for infant cries only; with males showing an increase on HRV, $M=10.43$, 95% CI [−12.95; 33.81], whereas females a decrease, $M=-17.25$, 95% CI [−37.84; 3.35] (see Fig. 5.1). When controlling for individual noise sensitivity, the Cue×Gender interaction effect was attenuated and became not significant, $F(1,21)=2.57$, $p=.124$, $\eta_p^2=.109$. Moreover, independently of the other factors, noise sensitivity was associated to a decrease of HRV, $F(1,21)=11.18$, $p=.003$, $\eta_p^2=.347$.

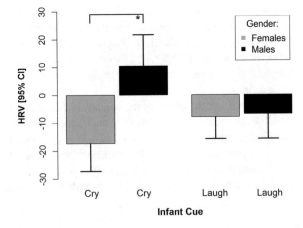

Fig. 5.1 Heart Rate Variability (HRV) as a function of the infant Cue (cry vs. laugh) and the Gender (males vs. females; $^*p<.05$)

5.3.2   Implicit Associations

The *ANOVA* conducted on the SC-IAT-As scores showed that implicit responses were not influenced by the Gender, $F < 1$, or the Cue$\times$Gender interaction, $F < 1$. Only the Cue main effect showed a tendency to significance, $F(1,23) = 3.51$, $p = .074$, $\eta_p^2 = .132$. The same results were confirmed also when controlling for individual noise sensitivity. Moreover, data showed that independently of the other factors, noise sensitivity was negatively associated with implicit evaluations, $F(1,22) = 6.61$, $p = .017$, $\eta_p^2 = .231$.

5.3.3   Explicit Responses

The *ANOVA* conducted on explicit responses showed that SD scores were influenced by the Gender, $F(1,23) = 4.82$, $p = 0.038$, $\eta_p^2 = 0.173$. The mean comparison showed that, females self-reported to have more positive explicit evaluations, $M = 6.31$, 95% CI [6; 6.62], than males, $M = 5.82$, 95% CI [5.47; 6.17]. The same result was confirmed also when controlling for individual noise sensitivity. Moreover, data showed that independently of the gender, noise sensitivity was negatively associated with explicit evaluations, $F(1,22) = 4.46$, $p = .046$, $\eta_p^2 = .169$.

5.3.4   Consistency of Responses

The correlation analysis showed that psychophysiological (HRV) and implicit responses (SC-IAT-A) to infant cries were positively associated, $r = .376$, $p = .032$, $N = 25$ (see Table 5.1). Therefore, the higher was the HR acceleration the more positive was the implicit attitude. Moreover, data showed that psychophysiological responses (HRV) to infant cries and laughs were positively associated, $r = .552$, $p = .002$, $N = 25$. No other significant association were observed.

   The same pattern of results was also observed when controlling by gender and individual anxiety (see Table 5.1).

5.4   Discussion and Conclusions

The present study investigated adults' reaction to salient infant cues by consider-ing different levels of processing, and investigating their consistency. The literature showed that individual responsiveness to infant signals is related to the quality of adult-infant interaction and associated with infant later development [1, 5, 7–9]. Fur-thermore, because noise sensitivity influences sound perception and evaluations, and

Table 5.1 Pearson and partial correlation coefficients between considered variable as a function of the levels of processing and the infant cue

Variables°	1	2	3	4	5
HRV					
1. Cries	–	.638***	.439*	.310	−.080
2. Laughs	.552**	–	.297	.274	−.062
SC-IAT-A					
3. Cries	.376*	.139	–	.153	−.002
4. Laughs	.322	.141	.242	–	−.243
Semantic differential					
5. Baby	−.142	−.111	.185	−.072	–

Note: °*HRV* heart rate variability; *SC-IAT-A* Single Category Implicit Association Test; *$p < .05$; **$p < .01$; ***$p < .001$; partial correlation (by gender and anxiety) are presented above the diagonal

that robust gender differences were observed on this dimension [22, 25], the individual noise sensitivity was considered as controlling factor. In line with the literature [18, 19, 21] (but see [16, 17] for a different result), we hypothesized that females would show a specific pattern of physiological and explicit responses, but not for implicit associations [22]; and a consistency of responses across the levels of processing. Moreover, we expected that gender differences were attenuated when the noise sensitivity was considered [22].

To test these hypotheses, two infant cues (cry and laugh) were presented, while participants' physiologic variations (HRV) and implicit associations (SC-IAT-A) were recorded; moreover, explicit evaluations (SD) toward babies were also collected. Different methodologies were used to take into account different levels of processing the infant cues. Indeed, parental models [5, 10] showed that adults' responses to infant cues are regulated at different levels, from more reflexive processes to more controlled ones. This is the first study that considers simultaneously three levels of infant cues processing, and investigates their associations.

In line with the literature [18, 19, 21], results showed that both cry and laugh were associated with variations in HRV, and confirmed that gender differences were observed for physiological and explicit responses only. No gender differences were observed on implicit associations [22]. As regards the HRV, data showed a decrease of HRV in females, whereas an increase in males, but only for the infant cry. No gender differences were observed for physiological responses related to infant laugh. As regards the explicit evaluations, results confirmed that women reported more positive evaluations toward babies than males. In sum, these results confirm that infant cues have a specific salience for all adults, and seems to indicate that the differences in males and females are mainly related to physiologic response and conscious or controlled evaluations, not to the implicit valence of the cue.

When controlling for WNSS, no gender differences were observed, but the noise sensitivity was associated with adults' responsiveness to infant cues. Higher noise sensitivity scores were associated to greater HRV decelerations, more negative implicit associations, and less positive explicit evaluations. Therefore, in line with previous studies [22, 25], data confirm the relevance of noise sensitivity on sound perception, and suggest that the observed gender effect could be also the expression of a general difference on noise sensitivity. Further studies are needed to investigate to what extent the gender differences are observed when controlling for individual noise sensitivity.

Finally, the analysis of the consistency of responses across levels showed that, independently of gender and anxiety, the physiological responses to the cues were strongly associated, and that physiological and implicit responses were consistent, but only for infant cry. Explicit evaluations were not associated to both physiological responses and implicit associations. Contrarily to our expectations, only a weak consistency of responses across the different levels of processing was observed. These results further confirm that adults' responsiveness to infant cues is a complex and multifactorial phenomenon [1, 5, 10].

Besides the merits of this study, some limitations should also be mentioned. First, the sample size was small, and this might have limited the statistical validity of the analyses. Future studies with bigger samples should replicate these findings. Second, we sampled only non-parents to investigate caregiving propensity independently of parental status. Further studies should apply multilevel methodology to compare responses to different infant cues across parents and non-parents. Finally, we considered adults' responses as an index of caregiving propensity, but no direct measure of caregiving was considered. Further study should include a direct measure of caregiving to investigate the predictive validity of the multilevel approach.

In conclusion, by means of a multilevel design, this study showed that adults' responses to infant cues are only partially consistent; thus further confirming that processes that regulate caregiving propensity are complex and multifactorial. In line with the recent literature, we believe that only an integrated multilevel approach could allow a deeper comprehension of adult-infant interactions and definition of optimal preventive interventions.

Acknowledgements The authors thank Maria Cristina Forte for her assistance in collecting the data for this study.

References

1. Bornstein, M.H.: Determinants of parenting. Dev. Psychopathol. Four **5**, 1–91 (2016). https://doi.org/10.1002/9781119125556
2. Zeifman, D.M.: An ethological analysis of human infant crying: answering Tinbergen's four questions. Dev. Psychobiol. **39**(4), 265–285 (2001). https://doi.org/10.1002/dev.1005
3. Beck, J.E., Shaw, D.S.: The influence of perinatal complications and environmental adversity on boys' antisocial behaviour. J. Child Psychol. Psychiatr. **46**, 35–46 (2005)

4. Putnick, D.L., Bornstein, M.H., Lansford, J.E., Chang, L., Deater-Deckard, K., Di Giunta, L., Bombi, A.S.: Agreement in mother and father acceptance-rejection, warmth, and hostility/rejection/neglect of children across nine countries. Cross Cult. Res. **46**, 191–223 (2012). https://doi.org/10.1177/1069397112440931
5. Barrett, J., Fleming, A.: Annual research review: all mothers are not created equal: neural and psychobiological perspectives on mothering and the importance of individual differences. J. Child Psychol. Psychiatry **52**(4), 368–397 (2011)
6. Belsky, J.: The determinants of parenting: a process model. Child Dev. **55**, 83–96 (1984). https://doi.org/10.2307/1129836
7. Kim, P., Feldman, R., Mayes, L.C., Eicher, V., Thompson, N., Leckman, J.F., Swain, J.E.: Breastfeeding, brain activation to own infant cry, and maternal sensitivity. J. Child Psychol. Psychiatry **52**, 907–915 (2011). https://doi.org/10.1111/j.1469-7610.2011.02406.x
8. Leerkes, E.M., Parade, S.H., Gudmundson, J.A.: Mothers' emotional reactions to crying pose risk for subsequent attachment insecurity. J. Fam. Psychol. **25**(5), 635–643 (2011)
9. McElwain, N.L., Booth-Laforce, C.: Maternal sensitivity to infant distress and nondistress as predictors of infant-mother attachment security. J. Fam. Psychol. **20**(2), 247–255 (2006)
10. Swain, J.E., Kim, P., Spice, J., Ho, S.S., Dayton, C.J., Elmadih, A., Abel, K.M.: Approaching the biology of human parental attachment: brain imaging, oxytocin and coordinated assessments of mothers and fathers. Brain Res. **1580**, 78–101 (2014)
11. Seifritz, E., Esposito, F., Neuhoff, J.G., Luthi, A., Mustovic, H., Dammann, G., von Bardeleben, U., Radue, E.W., Cirillo, S., Tedeschi, G., Di Salle, F.: Differential sex-independent amygdala response to infant crying and laughing in parents versus nonparents. Biol. Psychiatry **54**, 1367–1375 (2003)
12. De Pisapia, N., Bornstein, M.H., Rigo, P., Esposito, G., De Falco, S., Venuti, P.: Gender differences in directional brain responses to infant hunger cries. NeuroReport **243**(3), 142–146 (2013). https://doi.org/10.1097/WNR.0b013e32835df4fa
13. Frodi, A.M., Lamb, M.E.: Child abusers' response to infant smiles and cries. Child Dev. **51**(1), 238–241 (1980). https://doi.org/10.2307/1129612
14. Del Vecchio, T., Walter, A., O'Leary, S.G.: Affective and physiological factors predicting maternal response to infant crying. Infant Behav. Dev. **32**, 117–122 (2009)
15. Joosen, K.J., Mesman, J., Bakermans-Kranenburg, M.J., Pieper, S., Zeskind, P.S., van Ijzendoorn, M.H.: Physiological reactivity to infant crying and observed maternal sensitivity. Infancy **18**, 414–431 (2013). https://doi.org/10.1111/j.1532-7078.2012.00122.x
16. Anderson-Carter, I., Beroza, A., Crain, A., Gubernick, C., Ranum, E., Vitek R (2015) Differences between non-parental male and female response to infant crying. JASS (2015)
17. Cohen-Bendahan, C.C.C., van Doornen, L.J.P., De Weerth, C.: Young adults' reaction to infant crying. Infant Behav. Dev. **37**(1), 33–43 (2014)
18. Out, D., Pieper, S., Bakermans-Kranenburg, M.J., van Ijezendoorn, M.H.: Physiological reactivity to infant crying: a behavioral genetic study. Genes Brain Behav. **9**(8), 868–876 (2010). https://doi.org/10.1111/j.1601-183X.2010.00624.x
19. Brewster, A.L., Nelson, J.P., McCanne, T.L., Luca, D.R., Milner, J.S.: Gender differences in physiological reactivity to infant cries and smiles in military families. Child Abuse Negl. **22**(8), 775–788 (1998). https://doi.org/10.1016/S0145-2134(98)00055-6
20. Frodi, A.M., Lamb, M.E.: Fathers' and mothers' responses to infant smiles and cries. Infant Behav. Dev. **1**, 187–198 (1978). https://doi.org/10.1016/S0163-6383(78)80029-0
21. Boukydis, C.F., Burgess, R.L.: Adult physiological response to infant cries: effects of temperament of infant, parental status, and gender. Child Dev. **53**(5), 1291–1298 (1982)
22. Senese, V.P., Venuti, P., Giordano, F., Napolitano, M., Esposito, G., Bornstein, M.H.: Adults' implicit associations to infant positive and negative acoustic cues: moderation by empathy and gender. Q. J. Exp. Psychol. (2016)
23. Senese, V.P., De Falco, S., Bornstein, M.H., Caria, A., Buffolino, S., Venuti, P.: Human infant faces provoke implicit positive affective responses in parents and non-parents alike. PLoS ONE (2013). https://doi.org/10.1371/journal.pone.0080379

24. Dimitriev, D.A., Saperova, E.V., Dimitriev, A.D.: State anxiety and nonlinear dynamics of heart rate variability in students. PLoS ONE **11**(1), e0146131 (2016)
25. Senese, V.P., Ruotolo, F., Ruggiero, G., Iachini, T.: The Italian version of the noise sensitivity scale: measurement invariance across age, sex, and context. Eur. J. Psychol. Assess. **28**, 118–124 (2012). https://doi.org/10.1027/1015-5759/a000099
26. Kaufmann, T., Sutterlin, S., Shulz, S.M., Vogele, C.: ARTiiFACT: a tool for heart rate artifact processing and heart rate variability analysis. Behav. Res. Methods **43**(4), 1161–1170 (2011). https://doi.org/10.3758/s13428-011-0107-7
27. Greenwald, A.G., Nosek, B.A., Banaji, M.R.: Understanding and using the implicit association test: I. An improved scoring algorithm. J. Pers. Soc. Psychol. **85**(2), 197–216 (2003). https://doi.org/10.1037/a0015575
28. Osgood, C.E., Suci, G.C., Tannenbaum, P.H.: The Measurement of Meaning. University of Illinois Press, Urbana, IL (1957)
29. Spielberger, C.D., Gorsuch, R.L., Lushene, R., Vagg, P.R., Jacobs, G.A.: Manual for the State-Trait Anxiety Inventory. Consulting Psychologists Press, Palo Alto, CA (1983)

Chapter 6
Olfactory and Haptic Crossmodal Perception in a Visual Recognition Task

S. Invitto, A. Calcagnì, M. de Tommaso and Anna Esposito

Abstract Olfactory perception is affected by cross-modal interactions between different senses. However, although the effect of cross-modal interactions for smell have been well investigated, little attention has been paid to the facilitation expressed by haptic interactions with a manipulation of the odorous object's shape. The aim of this research is to investigate whether there is a cortical modulation in a visual recognition task if the stimulus is processed through an odorous cross-modal pathway or by haptic manipulation, and how these interactions may have an influence on early visual-recognition patterns. Ten healthy non-smoking subjects (25 years ± 5 years) were trained to have a haptic manipulation of 3-D models and olfactory stimulation. Subsequently, a visual recognition task was performed during an electroencephalography recording to investigate the P3 Event Related Potentials components. The subjects had to respond on the keyboard according to their subjective predominant recognition (olfactory or haptic). The effects of haptic and olfactory condition were assessed via linear mixed-effects models (LMMs) of the lme4 package. This model allows for the variance related to random factors to be controlled without any data aggregation. The main results highlighted that P3 increased in the olfactory cross-modal condition, with a significant two-way interaction between odor and left-sided

S. Invitto (✉)
Human Anatomy and Neuroscience Lab, Department of Environmental
Science and Technology, University of Salento, Lecce, Italy
e-mail: sara.invitto@unisalento.it

A. Calcagnì
Department of Psychology and Cognitive Science, University of Trento, Trento, Italy

M. de Tommaso
Department of Medical Science, Neuroscience, and Sense Organs,
University Aldo Moro, Bari, Italy

A. Esposito
Department of Psychology, University of Campania 'Luigi Vanvitelli',
Caserta, Italy

A. Esposito
IIASS, Vietri Sul Mare, Italy

© Springer International Publishing AG, part of Springer Nature 2019
A. Esposito et al. (eds.), *Quantifying and Processing Biomedical and Behavioral
Signals*, Smart Innovation, Systems and Technologies 103,
https://doi.org/10.1007/978-3-319-95095-2_6

lateralization. Furthermore, our results could be interpreted according to ventral and dorsal pathways as favorite ways to olfactory crossmodal perception.

Keywords Olfactory perception · Cross-modal perception · Haptic stimulation 3D shapes · Smell · P3

6.1 Introduction

The connectivity of brain sensory areas with other sensory modalities allows the integration of olfactory information with other sensory channels, which is the origination of multisensory and cross-modal perceptions [9, 15, 16].

The mechanisms by which different smells cause different brain responses have been described by the olfactory map model [7]. Olfactory receptors respond differently and systematically to the molecular features of odors. These features are encoded by neural activity patterns in the glomerular layer, which seem to create images representing odors. Such olfactory images play a role in the representation of perceived odors. The odor images are processed successively by microcircuits to provide the basis for the detection and discrimination of smells. The odor images, combined with taste, vision, hearing, and motor manipulation, provide the basis for the perception of flavors [26]. Such a complex reality reflects the need for the brain to develop strategies to quickly and effectively codify different sensorial inputs originating from different sensorial modalities.

Evidence of cross-modal activation in the olfactory system is highlighted by observing olfactory clinical dysfunctions, for example, anosmia. The most common symptom of anosmia is an interference with feeding; anosmic patients are not able to regularly detect the taste and smell of foods. In these cases, the patient may favor other senses.

The smell of food seems to be affected by tactile perception in young subjects and by visual impact in elderly subjects [18]. Experimental evidence investigating the cross-modal association between taste and vision showed that pleasant and unpleasant tastes are associated with round and angular shapes, respectively [9, 19].

Emotion-based information processing involves specific regions of the brain (insula and amygdala) that interact with the olfactory system. The amygdala then modulates the perception of facial expressions that describe specific emotional states. In fact, emotional face recognition is not merely a visual mechanism because it works through the integration of different multisensorial information (that is, voice, posture, social situations, and odors). In particular, the olfactory system appears highly involved in processing information about social interactions [16].

Although a large body of research exists on cross-modal interactions between olfaction and other senses, it has only been in the last decade that there has been a rise in the number of studies investigating the nature of multisensorial interactions between olfaction and touch. Results from our search of the literature highlighted that olfactory perception could modulate haptic perception in terms of different tactile dimensions, such as texture and temperature [5, 6]. Recent findings have shown that haptic and visual recognition are influenced in the same way by the orientation and size of objects [4].

A relevant role in cross-modal interaction is played by a subject's bias, either in naming or representing odors in mental imagery, even when not expressly required by the task. However, little attention has been paid to the facilitating role played by haptic interaction when manipulating the shape of an 'odorous' object.

The aim of the present research is to investigate whether there is cortical modulation during a visual recognition task if the stimulus, which represents a odorous object, is processed by an odorous cross-modal pathway or by haptic manipulation, and how these interactions may have an influence on early visual-recognition patterns. We investigate crossmodal perception through P3 Event Related Potential (ERP). P3 is an ERP component that we can be elicited through an odd ball task in an electroencephalic recording [17, 25]. Furthermore, P3 could be elicited in Olfactory task [20–22]. Moreover, although P3 is a non-specific component for the selection of shape patterns, it is a very sensitive component in the processing capacity [13], therefore the most suitable for this cross-modal protocol.

6.2   Method

6.2.1   Participants

Ten healthy non-smoking subjects (25 years ± 5 years) were recruited from the student population at the University of Salento to participate in the study. All subjects had normal smelling abilities and with normal vision. There were no reported current or past psychopathologies, neurological illnesses, or substance abuse problems.

Participants were instructed to not use perfume or drink coffee on the day of the test. Event-Related Potentials recording sessions were scheduled between 9 and 5 p.m. Each session had a duration of 1 h.

The experimental protocol was approved by the Ethical Committee of ASL (Local Health Company) of Lecce, and informed consent was obtained from participants according to the Helsinki Declaration.

6.2.2   Stimuli and Procedure

We arranged an experiment of olfactory stimulation by analyzing visual ERPs after a training of cross-modal haptic and olfactory interaction with nine diluted odorants and three-dimensional (3-D) shapes (Fig. 6.1). The smells were selected from five representative types of categorical spatial dimensions [12]. Six odors were presented in Cross-modal olfactory and haptic condition (i.e., Lemon, Cinnamon, Mushrooms, Banana, Grass, Eucalyptol) were presented in olfactory crossmodal haptic mode, 5 odors were presented in olfactory condition (i.e., mint, Rose, Geraniol, Almond, Flowers) and 27 odors were presented for the olfactory visual condition (e.g., Apple,

Fig. 6.1 Example of three-dimensional haptic shapes printed for the experiment

Go/No-Go Task

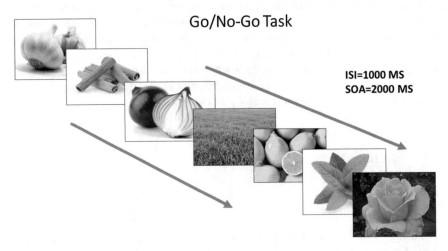

ISI=1000 MS
SOA=2000 MS

Fig. 6.2 Example of the Go/No-Go Task presented during the experiment. The instructions were: "Please, press the left-hand side button if your predominant recognition of the stimulus has been encoded through olfactory stimulation and the right-hand side button if it has been encoded through haptic stimulation"

Chocolate, Potato, Salt, Garlic, fishing, vanilla, strawberries, rice, salt, pasta, poop, Lemon, Cinnamon, Mushrooms, Banana, Grass, Eucalyptol, Mint, Rose, Geraniol, Almond, Flowers and so on).

Subjects were trained to have haptic manipulation of 3-D models (which were created using the 3-D Blender 2.74 platform) and olfactory stimulation in a black case through an olfactory stimulation device [11]. Stimulations were presented in a blind modality (the subject didn't have any visual information about the odorant or about the shapes). Each stimulation had a duration of 1000 ms.

After the training, subjects had to perform a computer-based visual recognition task. During the task, two-dimensional (2D) visual stimuli (a repertoire of images that represented edible and scented substances) were presented to the subjects. The images of the Go/No-Go Task were presented using the software ePrime 2.0 (Fig. 6.2). During the Go/No-Go Task, the interstimulus interval (ISI) had a duration of 1000 ms, the stimulus presentation had a duration of 1000 ms, and the stimulus-onset asynchrony (SOA) had a duration of 2000 ms. An EEG recording (64-channel actiCHamp, Brain Products) was made during the task for each subject.

During the 2-D session, the subjects were tasked with pressing a button on the left-hand side of the keyboard if the predominant recognition of the stimulus had been encoded through olfactory modality and a button on the right-hand side of the keyboard if the predominant recognition of the stimulus has been encoded in haptic modality. After the Go/No-Go Task a Visual Analogic Scale (VAS) was administered to the subjects to investigate the pleasantness, the level of arousal and the familiarity of the different conditions.

6.3   Data Analysis

Statistical analyses were performed using linear mixed-effects models (LMMs) and the lme4 package [2], which were available through the R Project for Statistical Computing program (version 3.1.1). Unlike traditional analyses of repeated measures, LMMs allow for analyses of unbalanced datasets and simultaneous estimation of group (fixed) and individual (random) effects [23] without averaging across trials. These kinds of statistical models are becoming popular in psychophysiology over the last decades [28]. In the current study, separate LMMs were run to evaluate the effect of the conditions (odor and haptic vs. visual condition) and lateralization (left, median, right) on amplitude and latency of the P3 ERP components. In each model, we considered the condition and the lateralization as fixed effects, and participant variability was coded as a random effect. The interaction between the condition and the lateralization was also checked. Results are described by assessing fixed effects in terms of beta coefficients of regressors (βs), standard errors (SEs), and t-values (Ts). Due to the distributional characteristics of the variables used in this study, models were estimated using the DAS-robust algorithm and implemented using the rlmer function in the R package [14]. In the context of robust-LMMs, significant effects were detected using the decision rule | t | > 2.0 because there were no common ways to compute degrees of freedom and, relatedly, p-values of regressors [1].

6.4   Results

Behavioural Results: Descriptive statistics values of VAS dimensions are described in Table 6.1. Table 6.1 indicates that crossmodal condition is valued as more pleasant, more arousing and more familiar than Visual condition.

During the Go/No-Go task the subjects respond with the same proportion to their stimulus encoding (51% olfactory encoding; 49% haptic encoding) (Fig. 6.3).

Descriptive value of Reaction Time Response (RTR) indicated a faster mean RTR in Olfactory Encoding (909.43 ms; SD = 39.30) than in Haptic Encoding (1046.84 ms; SD = 71.32) (Fig. 6.4).

Psychophysiological Results: Table 6.2 shows results for the Amplitude component of P3. They revealed a significant effect of Condition ($\chi^2_2 = 79.27$, $p < .001$),

	Haptic and odor condition	Visual condition
Pleasantness	3.37 (1.41)	2.60 (1.68)
Arousing	3.69 (1.12)	3.40 (0.51)
Familiar	4.35 (0.78)	3.20 (1.20)

Table 6.1 Descriptive statistics of VAS dimensions: mean values and standard deviations

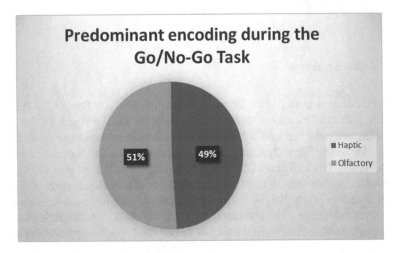

Fig. 6.3 Proportion of predominant encoding during the Go/No-Go task

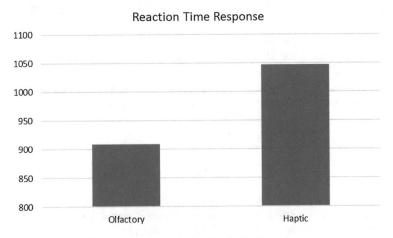

Fig. 6.4 Behavioral reaction time response during the Go/No-Go task

Lateralization ($\chi_1^2 = 5.99$, $p = .01$), and Localization ($\chi_4^2 = 61.04$, $p < .001$) as well as a significant interaction of Condition $\times$ Localization ($\chi_8^2 = 20.89$, $p = .002$) on P3 Amplitude. Particularly, a more positive P3 waveform was observed in the Smell-condition (B $= 1.84$, $t_{1289} = 2.05$, $p = .03$), in the left-side lateralization

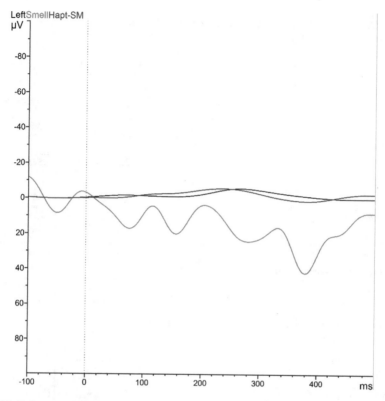

Fig. 6.5 Left comparison: cross-modal smell condition (red line); haptic condition (blue line); visual condition (black line)

(B = 2.22, t_{1287} = 2.59, p = .009) (Fig. 6.5), and in the Parietal area (B = 2.35, t_{1287} = 2.29, p = .02) (Fig. 6.6). Moreover, positive waveforms of P3 were also found in the Central area during Haptic manipulation (B = 2.80, t_{1287} = 2.18, p = .02) (see Loreta source reconstruction for Haptic Condition Fig. 6.7). Similarly, the left-side Temporal region over the Smell condition revealed a positive P3 Amplitude (B = 5.30, t_{1289} = 2.99, p = .002) (see Loreta source reconstruction for Smell Condition Fig. 6.8) respect Visual Condition (Fig. 6.9).

Table 6.3 shows instead results for the Latency component of P3. Significant effects were found for Condition (χ^2_2 = 21.28, p < .001), Localization (χ^2_4 = 48.80, p < .001), and for the interaction Localization × Lateralization (χ^2_4 = 12.27, p = .01). Particularly, latency increased in the Left-side lateralization (B = 21.79, t_{1812} = 3.28, p = .001) as well as in Central (B = 42.39, t_{1287} = 5.83, p < .001), Occipital (B = 27.19, t_{1287} = 3.17, p = .001), Parietal (B = 30.95, t_{1287} = 3.97, p < .001), and Temporal (B = 15.75, t_{1287} = 2.37, p = .01) areas. On the contrary, latency decreased in the Central area during Haptic manipulation (B = −27.24, t_{1812} = −2.65, p = .008)

Table 6.2 Results of linear mixed-effects model: fixed effects for manipulation, lateralization, and localization on amplitude (P3)

	χ^2(df)	B(SE)	t
Baseline		0.264(0.902)	0.293
Condition	79.27(2)***		
Smell versus non-smell		1.8480.8979)	2.059*
Haptic versus non-smell		−0.514(0.8831)	−0.583
Lateralization	5.99(1)*		
Left-side versus right-side		2.224(0.857)	2.595**
Localization	61.04(4)***		
Central (C)		1.417(0.9033)	1.569
Occipital (O)		1.172(1.1593)	1.011
Parietal (P)		2.3596(1.0285)	2.294*
Temporal (T)		0.913(0.8578)	1.065
Condition × Lateralization	0.09(2)		
Smell × Left-side		−2.298(1.2349)	−1.862
Haptic × Left-side		−1.417(1.2425)	−1.141
Condition × Localization	20.89(8)**		
Smell × Central (C)		2.001(1.3309)	1.504
Haptic × Central (C)		2.805(1.2869)	2.180*
Smell × Occipital (O)		0.272(1.748)	0.156
Haptic × Occipital (O)		0.699(1.6856)	0.415
Smell × Parietal (P)		2.114(1.55)	1.364
Haptic × Parietal (P)		1.584(1.483)	1.068
Smell × Temporal (T)		0.142(1.2549)	0.114
Haptic × Temporal (T)			−0.137
Lateralization × Localization	5.82(4)		
Left-side versus Central (C)		−1.837(1.2369)	−1.485
Left-side versus Occipital (O)		−2.206(1.6561)	−1.332
Left-side versus Parietal (P)		−2.633(1.4274)	−1.845
Left-side versus Temporal (T)		−2.136(1.1984)	−1.783
Condition × Lateralization × Localization	14.68(8)		
Smell × Left-side × Central (C)		2.336(1.8143)	1.288
Haptic × Left-side × Central (C)		0.403(1.7743)	0.227
Smell × Left-side × Occipital (O)		2.506(2.5448)	0.985
Haptic × Left-side × Occipital (O)		1.383(2.4531)	0.564
Smell × Left-side × Parietal (P)		2.028(2.1456)	0.945
Haptic × Left-side × Parietal (P)		2.586(2.0715)	1.248

(continued)

Table 6.2 (continued)

	χ^2(df)	B(SE)	t
Smell × Left-side × Temporal (T)		5.302(1.7715)	2.993**
Haptic × Left-side × Temporal (T)		2.689(1.7658)	1.523

Notes
Subjects were treated as random effects, degrees of freedom of the model were calculated with the Satterthwaite approximation. Reference levels for contrasts: Non-smell (Condition), Frontal (Localization), Right-side (Lateralization). Values of χ^2 are computed with the type-II Wald test. $N_{obs} = 1324$, $N_{groups} = 12$, $ICC_{groups} = 0.15$
* $p < .05$ ** $p < .01$ *** $p < .001$

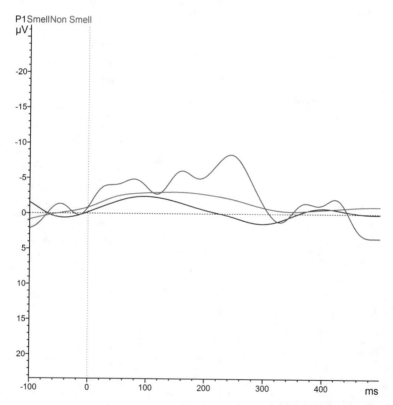

Fig. 6.6 Parietal left comparison: cross-modal smell condition (blue line); haptic condition (black line); visual condition (green line)

and in left-side of Central (B = −24.20, t_{1812} = 2.41, p = .01), Occipital (B = −32.43, t_{1812} = −2.67, p = .007), and Parietal (B = −22.98, t_{1812} = 2.10, p = .03).

This Areas is involved in different aspects of memory than the medial temporal lobes Retrograd memory.

Table 6.3 Results of linear mixed-effects model: fixed effects for manipulation, lateralization, and localization on Latency (P3)

	χ^2(df)	B(SE)	t
Baseline		271.274(6.369)	42.594
Condition	21.271(2)***		
Smell versus non-smell		11.551(6.794)	1.700
Haptic versus non-smell		6.714(6.636)	1.012
Lateralization	0.871(1)		
Left-side versus right-side		21.798(6.636)	3.285**
Localization	48.803(4)***		
Central (C)		42.393(7.269)	5.832***
Occipital (O)		27.198(8.567)	3.175**
Parietal (P)		30.955(7.781)	3.978***
Temporal (T)		15.750(6.636)	2.374*
Condition × Lateralization	1.683(2)		
Smell × Left-side		−15.733(9.595)	−1.640
Haptic × Left-side		−10.131(9.384)	−1.080
Condition × Localization	13.540(8)		
Smell × Central (C)		−17.445(10.511)	−1.660*
Haptic × Central (C)		−27.248(10.280)	−2.651**
Smell × Occipital (O)		−13.826(12.387)	−1.116
Haptic × Occipital (O)		−22.298(12.115)	−1.840
Smell × Parietal (P)		−9.894(11.251)	−0.879
Haptic × Parietal (P)		−15.464(11.004)	−1.405
Smell × Temporal (T)		1.406(9.595)	0.147
Haptic × Temporal (T)		−8.786(9.384)	−0.936
Lateralization × Localization	12.278(4)*		
Left-side versus Central (C)		−24.200(10.027)	−2.413*
Left-side versus Occipital (O)		−32.437(12.115)	−2.677**
Left-side versus Parietal (P)		−22.985(11.004)	−2.089*
Left-side versus Temporal (T)		−16.143(9.384)	−1.720
Condition × Lateralization × Localization	5.150(8)		
Smell × Left-side × Central (C)		13.984(14.499)	0.964
Haptic × Left-side × Central (C)		6.580(14.193)	0.464
Smell × Left-side × Occipital (O)		33.766(17.519)	1.927
Haptic × Left-side × Occipital (O)		24.437(17.133)	1.426
Smell × Left-side × Parietal (P)		15.966(15.912)	1.003
Haptic × Left-side × Parietal (P)		8.173(15.562)	0.525
Smell × Left-side × Temporal (T)		2.220(13.560)	0.164
Haptic × Left-side × Temporal (T)		0.071(13.271)	0.005

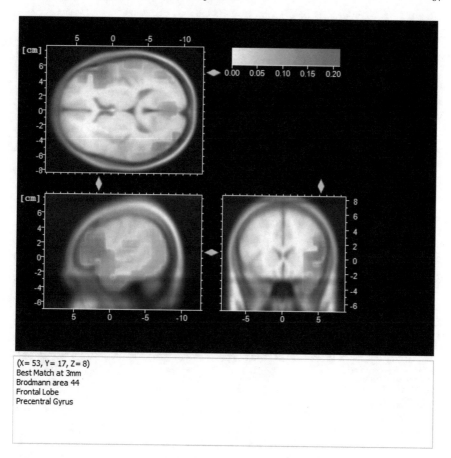

Fig. 6.7 Loreta Haptic condition. Brodmann area 44—this area involves the premotor functions

6.5   Discussion

The present research results highlight changes in the P3 ERP components. P3 is a perceptual and cognitive component of ERPs, that is related to stimulus detection. P3 is recorded in relation to familiar, but infrequent stimuli [24, 27]. ERP variations are evident in the odorous state and in the manipulation condition. In fact, as shown in Fig. 6.3, the condition of simply visual recognition in this paradigm does not produce an obvious P3 component, which is in fact considerably elicited in the odorous condition. The Odorous and Haptic condition is lateralized in the left hemisphere, particularly in the occipital, temporal, and parietal areas, which can be defined as 'occipital–temporal–parietal streams'. In addition to being particularly relevant because it is located in the left hemisphere, this area is where the "semantic" function of language and categorical perception reside [10]. These findings could also suggest a dorsal pathway on the visual path of localization known as 'how to do'

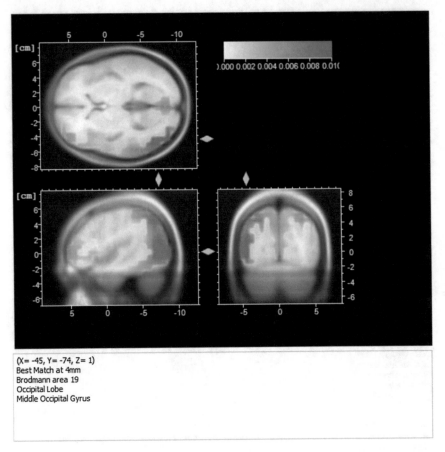

(X= -45, Y= -74, Z= 1)
Best Match at 4mm
Brodmann area 19
Occipital Lobe
Middle Occipital Gyrus

Fig. 6.8 Loreta smell condition: Brodmann area 19, that is activated by somatorosensory stimuli

[3, 8]. In this case, we could connect haptic and olfactory manipulation to the dorsal pathway (for example, "I smell and I manipulate a shape and, thus, I create imagery of haptic action"). Furthermore, we could, in part, link the visual condition to the 'ventral' pathway (that is, temporal and occipital locations), which is linked more to the representation of the imagery of the smelled object and which is then recognized in a visual mode Globally, we could interpret these results on the two components just as a predominant olfactory encoding in the crossmodal task, which is evident in the P3 ERP. Moreover, the activation, for the olfactory encoding component, of the Brodmann Area 19 (Fig. 6.8), area connected to somatorosensory stimuli and to the retrograde memory, seems particularly interesting. This seems to be precisely the key to understanding the greater amplitude and lateralization found in olfactory modality. The arousal in this case could be due to a greater stimulus processing that requires greater memory resources, and which allows on the one hand a wider potential, on the other hand, wider behavioral reaction times.

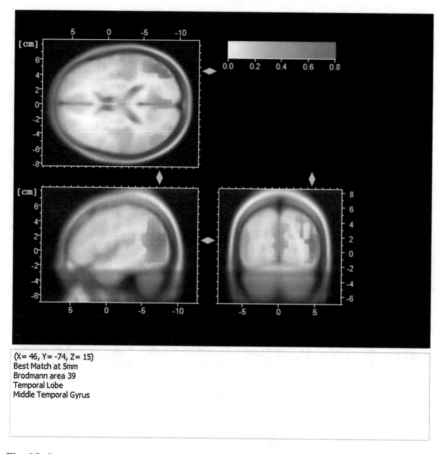

(X= 46, Y= -74, Z= 15)
Best Match at 5mm
Brodmann area 39
Temporal Lobe
Middle Temporal Gyrus

Fig. 6.9 Loreta non smell condition: Brodmann area 39. This area is involved in semantic memory

Acknowledgements We would like to thank to Graziano Scalinci, who printed the 3-D shapes; and Federica Basile and Francesca Tagliente, who collaborated with the EEG data recording. Paper co-funded through 5× Thousand Research Fund-University of Salento.

References

1. Baayen, R.H.: Analyzing Linguistic Data: A Practical Introduction to Statistics Using R. Cambridge University Press (2008). https://doi.org/10.1017/CBO9780511801686
2. Bates, D., Maechler, M., Bolker, B., Walker, S.: Package lme4. J. Stat. Softw. **67**(1), 1–91 (2015). http://lme4.r-forge.r-project.org
3. Bornkessel-Schlesewsky, I., Schlesewsky, M.: Reconciling time, space and function: a new dorsal-ventral stream model of sentence comprehension. Brain Lang. **125**(1), 60–76 (2013). https://doi.org/10.1016/j.bandl.2013.01.010

4. Craddock, M., Lawson, R.: Repetition priming and the haptic recognition of familiar and unfamiliar objects. Percept. Psychophys. **70**(7), 1350–1365 (2008). https://doi.org/10.3758/PP.70.7.1350
5. Dematte, M.L.: Cross-modal interactions between olfaction and touch. Chem. Senses **31**(4), 291–300 (2006). https://doi.org/10.1093/chemse/bjj031
6. Fernandes, A.M., Albuquerque, P.B.: Tactual perception: a review of experimental variables and procedures. Cogn. Process. **13**(4), 285–301 (2012). https://doi.org/10.1007/s10339-012-0443-2
7. Giessel, A.J., Datta, S.R.: Olfactory maps, circuits and computations. Curr. Opin. Neurobiol. (2014). https://doi.org/10.1016/j.conb.2013.09.010
8. Goodale, M.A., Króliczak, G., Westwood, D.A.: Dual routes to action: Contributions of the dorsal and ventral streams to adaptive behavior. In: Progress in Brain Research, vol. 149, pp. 269–283 (2005). http://doi.org/10.1016/S0079-6123(05)49019-6
9. Hanson-Vaux, G., Crisinel, A.S., Spence, C.: Smelling shapes: Crossmodal correspondences between odors and shapes. Chem. Senses **38**(2), 161–166 (2013). https://doi.org/10.1093/chemse/bjs087
10. Holmes, K.J., Wolff, P.: Does categorical perception in the left hemisphere depend on language? J. Exp. Psychol. Gen. **141**(3), 439–443 (2012). https://doi.org/10.1037/a0027289
11. Invitto, S., Capone, S., Montagna, G., Siciliano, P.A.: MI2014A001344 Method and system for measuring physiological parameters of a subject undergoing an olfactory stimulation (2014)
12. Jourdan, F.: Spatial dimension in olfactory coding: a representation of the 2-deoxyglucose patterns of glomerular labeling in the olfactory bulb. Brain Res. **240**(2), 341–344 (1982). https://doi.org/10.1016/0006-8993(82)90232-3
13. Kok, A.: On the utility of P3 amplitude as a measure of processing capacity. Psychophysiology **38**(3), 557–577 (2001). https://doi.org/10.1017/S0048577201990559
14. Kuznetsova, A., Brockhott, P.B., Christensen, R.H.B.: lmerTest: Tests in Linear Mixed Effects Models. R Package Version (2015). http://CRAN.R-project.org/package=lmerTest
15. Leleu, A., Demily, C., Franck, N., Durand, K., Schaal, B., Baudouin, J.Y.: The odor context facilitates the perception of low-intensity facial expressions of emotion. PLoS ONE **10**(9), 1–19 (2015). https://doi.org/10.1371/journal.pone.0138656
16. Leleu, A., Godard, O., Dollion, N., Durand, K., Schaal, B., Baudouin, J.Y.: Contextual odors modulate the visual processing of emotional facial expressions: An ERP study. Neuropsychologia **77**, 366–379 (2015). https://doi.org/10.1016/j.neuropsychologia.2015.09.014
17. Luck, S.J.: An Introduction to Event-related Potentials and Their Neural Origins. An Introduction to the Event-Related Potential Technique, 2–50 (2005)
18. Merkonidis, C., Grosse, F., Ninh, T., Hummel, C., Haehner, A., Hummel, T.: Characteristics of chemosensory disorders—results from a survey. Eur. Arch. Otorhinolaryngol. **272**(6), 1403–1416 (2014). https://doi.org/10.1007/s00405-014-3210-4
19. Ngo, M.K., Misra, R., Spence, C.: Assessing the shapes and speech sounds that people associate with chocolate samples varying in cocoa content. Food Qual. Prefer. **22**(6), 567–572 (2011). https://doi.org/10.1016/j.foodqual.2011.03.009
20. Nordin, S., Andersson, L., Olofsson, J.K., McCormack, M., Polich, J.: Evaluation of auditory, visual and olfactory event-related potentials for comparing interspersed- and single-stimulus paradigms. Int. J. Psychophysiol. **81**(3), 252–262 (2011). https://doi.org/10.1016/j.ijpsycho.2011.06.020
21. Pause, B.M., Krauel, K.: Chemosensory event-related potentials (CSERP) as a key to the psychology of odors. Int. J. Psychophysiol. (2000). https://doi.org/10.1016/S0167-8760(99)00105-1
22. Pause, B.M., Sojka, B., Krauel, K., Fehm-Wolfsdorf, G., Ferstl, R.: Olfactory information processing during the course of the menstrual cycle. Biol. Psychol. **44**(1), 31–54 (1996). https://doi.org/10.1016/S0301-0511(96)05207-6
23. Pinheiro, J.C., Bates, D.M.: Mixed Effects Models in S and S-Plus. Springer, New York (2000). http://doi.org/10.1198/tech.2001.s574

24. Polich, J., Criado, J.R.: Neuropsychology and neuropharmacology of P3a and P3b. Int. J. Psychophysiol. **60**(2), 172–185 (2006). https://doi.org/10.1016/j.ijpsycho.2005.12.012
25. Polich, J., Kok, A.: Cognitive and biological determinants of P300: an integrative review. Biol. Psychol. **41**(2), 103–146 (1995). https://doi.org/10.1016/0301-0511(95)05130-9
26. Shepherd, G.M.: Smell images and the flavour system in the human brain. Nature, **444**(7117), 316–321 (2006). Retrieved from http://www.ncbi.nlm.nih.gov/pubmed/17108956
27. Silverstein, B.H., Snodgrass, M., Shevrin, H., Kushwaha, R.: P3b, consciousness, and complex unconscious processing. Cortex; J. Devoted Study Nerv. Syst. Behav. **73**, 216–227 (2015). https://doi.org/10.1016/j.cortex.2015.09.004
28. Tremblay, A., Newman, A.J.: Modeling nonlinear relationships in ERP data using mixed-effects regression with R examples. Psychophysiology **52**(1), 124–139 (2015). https://doi.org/10.111 1/psyp.12299

Chapter 7
Handwriting and Drawing Features for Detecting Negative Moods

Gennaro Cordasco, Filomena Scibelli, Marcos Faundez-Zanuy, Laurence Likforman-Sulem and Anna Esposito

Abstract In order to provide support to the implementation of on-line and remote systems for the early detection of interactional disorders, this paper reports on the exploitation of handwriting and drawing features for detecting negative moods. The features are collected from depressed, stressed, and anxious subjects, assessed with DASS-42, and matched by age and gender with handwriting and drawing features of typically ones. Mixed ANOVA analyses, based on a binary categorization of the groups, reveal significant differences among features collected from subjects with negative moods with respect to the control group depending on the involved exercises and features categories (in time or frequency of the considered events). In addition, the paper reports the description of a large database of handwriting and drawing features collected from 240 subjects.

Keywords Handwriting · Depression–anxiety–stress scales (DASS)
Emotional state · Affective database

G. Cordasco (✉) · A. Esposito
Dipartimento di Psicologia, Universitá degli Studi della Campania "L. Vanvitelli", Caserta, Italy
e-mail: gennaro.cordasco@unicampania.it

G. Cordasco · A. Esposito
International Institute for Advanced Scientific Studies (IIAS), Vietri sul Mare, Italy
e-mail: iiass.annaesp@tin.it

F. Scibelli
Dipartimento di Studi Umanistici, Universitá degli Studi di Napoli "Federico II", Naples, Italy
e-mail: filomena.scibelli@unina.it

M. Faundez-Zanuy
Escola Superior Politecnica, TecnoCampus Mataro-Maresme, 08302 Mataro, Spain
e-mail: faundez@tecnocampus.cat

L. Likforman-Sulem
Télécom ParisTech, Université Paris-Saclay, 75013 Paris, France
e-mail: likforman@telecom-paristech.fr

© Springer International Publishing AG, part of Springer Nature 2019
A. Esposito et al. (eds.), *Quantifying and Processing Biomedical and Behavioral Signals*, Smart Innovation, Systems and Technologies 103,
https://doi.org/10.1007/978-3-319-95095-2_7

7.1   Introduction

Early detection of diseases is the key for health care. The earlier a disease is diagnosed, the more likely it is that it can be cured or successfully managed. When you treat a disease early, you may be able to prevent or delay problems from the disease.

Several chronic diseases are asymptomatic and can be diagnosed by special check-ups. Unfortunately, such checkups are costly and often invasive and therefore they are not chosen on the basis of possible trade-offs in terms of costs and benefits. On the other hand, the detection of illnesses can be also based on the analysis of simple non invasive human activities (speeches, body motions, physiological data, hand-writings, and drawings) [3, 12]. In particular, it has been shown that several diseases like Parkinson and Alzheimer have a sensible effect on writing skills [16].

Handwriting is a creative process which involves conscious and unconscious brain functionalities [15]. Indeed, handwriting and drawing analyses on paper has been suc-cessfully used for cognitive impairment detection and personality trait assessment [12, 14, 18, 25]. To this aim, several tests like the clock drawing test (CDT) [10], the mini mental state examination (MMSE) [9] and the house-tree-person (HTP) [11] have been designed in the medical domain. Recently, thanks to the development of novel technological tools (scanners, touch displays, display pens and graphic tablets), it is possible to perform handwriting analyses on computerized platforms. Such com-puterized analyses provide two main advantages: (i) the collection of data includes several non-visible features such as pressure, pen inclination, in air motions; (ii) data includes timing information which enable the *online*analysis the whole drawing and/or handwriting process instead of analyzing a single *offline* snapshot represent-ing the final drawing. Accordingly tests for detecting brain stroke risk factors and neuromuscular disorders, or dementia, have been computerized [10, 17, 19].

Accurate recognition of emotions is an important social skill that enable to prop-erly establish successful interactional exchanges [5, 6, 8, 22]. Indeed, emotions and moods affect cognitive processes, such as executive functions, and memory [20, 26] and have an effect on handwriting, since it involves visual, sensory, cognitive, and motor mechanisms.

The first study that applied handwriting analysis for the detection of emotions is [12]. It presents a public database (EMOTHAW) which relates emotional states to handwriting and drawing. EMOTHAW includes samples of 129 participants whose emotional states, namely anxiety, depression, and stress, are assessed by the Depression-Anxiety-Stress Scales (DASS-42) questionnaire [13]. Handwriting and drawing tasks have been recorded through a digitizing tablet. Records consist in pen positions, pen states (on-paper and in-air), time stamps, pen pressure and inclination. The paper reports also a preliminary analysis on the presented database, where sev-eral handwriting features, such as the time spent "in air" and "on paper", as well as, the number of strokes, have been identified and validated in order to detect anxiety, depression, and stress.

This paper extends the work in [12] presenting an enlarged database of 240 sub-jecta and proposing also some novel data-analysis features in order to measure the effects of negative moods on handwriting and drawing performances. Mixed ANOVA analyses, based on a binary categorization of the groups (i.e., typical, stressed, anxious, and depressed subjects), reveal significant differences among features collected from subjects with negative moods with respect to the control group depending on the involved exercises and features categories (in time or frequency of the considered events).

7.2 Handwriting Analysis for the Detection of Negative Moods

7.2.1 Handwriting Analysis

Handwriting analysis have been carried out using an INTUOS WACO series 4 digitizing tablet and a special writing device named Intuos Inkpen. Participants were required to write on a sheet of paper (DIN A4 normal paper) laid on the tablet. Figure 7.1 shows the sample acquired from one participant. While the digital signal is visualized on the screen, it is also visible on the paper (due to the inkpen), and the participant normally looks at the paper. There is a human supervisor sitting next to the participant, controlling a specifically designed acquisition software linked to the tablet.

For each task, the software starts automatically to capture data as soon as the inkpen touches the paper. The recording is stopped manually by the supervisor but this does not affect the collected timing information since the software does not record any data while the pen is far from the tablet. The software stores the data in a svc file (a simple ASCII file that can be opened with any text editor).

During each experimental task, the following information is continuously captured with a frequency of 125 Hz (see also Fig. 7.2):

1. position in x-axis;
2. position in y-axis;
3. time stamp;
4. pen status (up = 0 or down = 1);
5. azimuth angle of the pen with respect to the tablet;
6. altitude angle of the pen with respect to the tablet;
7. pressure applied by the pen on the paper.

Using this set of dynamic data, further information such as acceleration, velocity, instantaneous trajectory angle, instantaneous displacement, time features, and ductus-based features (see Sect. 7.4.2.2) can be inferred. The system has the nice property to capture in-air movements (when the inkpen is very close to the paper) which are lost using the on-paper ink. The in-air information has proven to be as important as the on-surface information [12, 21, 23]. In particular, in [12] several

Fig. 7.1 A4 sheet with the
set of tasks filled by one
participant

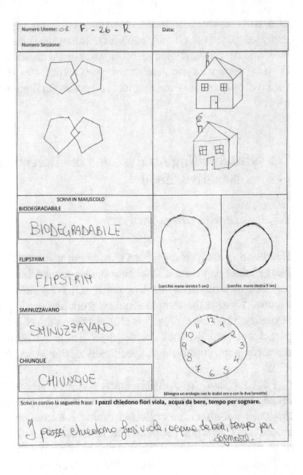

simple features exploited this in air (up) or on paper (down) information. In this
paper we performed a more detailed classification by partitioning the pen states into
three categories:

- **up**, recorded with state 0;
- **down**, recorded with state 1;
- **idle**, not recorded but recognizable using time stamps. Usually the tablet collects
  information with a frequency of 125 Hz (one row each 0.008 s). When two consecutive row are separated by more than 0.008 s the missing time is considered as
  an idle state.

The tasks acquired by the tablet are the following (see Fig. 7.1):

1. copy of a two-pentagon drawing;
2. copy of a house drawing;

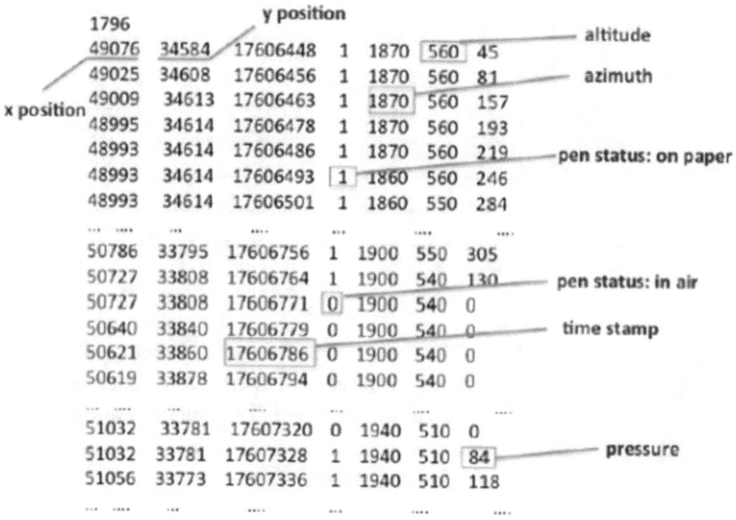

```
         1796              y position
         49076  34584  17606448  1  1870  560  45  ──────── altitude
         49025  34608  17606456  1  1870  560  81  ──────── azimuth
         49009  34613  17606463  1  1870  560  157
x position 48995  34614  17606478  1  1870  560  193
         48993  34614  17606486  1  1870  560  219 ──── pen status: on paper
         48993  34614  17606493  1  1860  560  246
         48993  34614  17606501  1  1860  550  284

         ...  ....   ...   .....   ...   ....   ....
         50786  33795  17606756  1  1900  550  305
         50727  33808  17606764  1  1900  540  130 ──── pen status: in air
         50727  33808  17606771  0  1900  540  0
         50640  33840  17606779  0  1900  540  0  ──── time stamp
         50621  33860  17606786  0  1900  540  0
         50619  33878  17606794  0  1900  540  0

         ...  ....   ...   .....   ...   ....   ....
         51032  33781  17607320  0  1940  510  0
         51032  33781  17607328  1  1940  510  84 ──────── pressure
         51056  33773  17607336  1  1940  510  118

         ...  ....   ...   .....   ...   ....   ....
```

Fig. 7.2 Extract of an svc file

3. writing of four Italian words in capital letters (BIODEGRADABILE (biodegradable), FLIPSTRIM (flipstrim), SMINUZZAVANO (to crumble), CHIUNQUE (anyone));
4. loops with left hand;
5. loops with right hand;
6. clock drawing;
7. writing of the following phonetically complete Italian sentence in cursive letters (I pazzi chiedono fiori viola, acqua da bere, tempo per sognare: Crazy people are seeking for purple flowers, drinking water and dreaming time).

7.2.2 Negative Moods: Depression, Anxiety and Stress Scales

In order to measure the subject current mood, the Italian version [24] of *DASS-42* (Depression, Anxiety and Stress Scales) [13] was administered to each participant. This is a self-report questionnaire composed by 42 statements measuring depressive, anxious, and stress symptoms. Each scale (D Depression; A Anxiety; S Stress) is made up by 14 statements. The scale is based on the assumption that depressed, anxyous, and stressed mood are not categorical (that is disorders), but dimensional constructs [13]. Hence, the mood can change from normal to severe state along a continuum. Psychometric properties of this scale were originally assessed in [13] on a large non-clinical sample (2.914), composed mostly by university students, identifying mood (depressed, anxious, stressed) severity ratings from normal to extremely severe, as showed by Table 7.1. Later, several studies assessed DASS psychometric

Table 7.1 DASS-42: severity rating [13]

	Depression (D)	Anxiety (A)	Stress (S)
Normal	0–9	0–7	0–14
Mild	10–13	8–9	15–18
Moderate	14–20	10–14	19–25
Severe	21–27	15–19	26–33
Extremely severe	28+	20+	34+

properties (competing models of the structure, reliabilities, convergent and discriminant validity) [1, 2, 4, 7]. The DASS administration procedure requires that for each statement in the questionnaire, the subject is asked "how much it applies to him/her over the past week". Participant answers are rated on a 4 points Likert scale (0 = never, 1 = sometimes; 2 = often; 3 = almost always).

7.3   The Database

For the data collection, 240 subjects ($126m$, $114f$) between 18 and 32 years ($M = 24.58$; $SD = \pm 2.47$) were recruited at the Department of Psychology of the Università degli Studi della Campania "L. Vanvitelli" (Caserta, Italy). All subjects were Master or BS students and volunteered their participation to the study. Participants were individually tested in a laboratory free of auditory and/or visual disturbances. At the beginning of the experiment, the study has been explained to the participant and, subsequently, he signed an informed consent. Before administering the DASS through a computer aided procedure and collecting handwriting samples, participants were informed that they can withdraw their data at any point of the data collection process. In addition, once collected, data were automatically anonymized in a way that experimenters would not be able later to identify participant's identities. Each participant first completed the *DASS-42* questionnaire and then the handwriting tasks. Table 7.2 shows the percentage of co-occurrence of emotional states resulting by DASS scores.

Aggregating the data it can be observed that 44.6% of subjects report a normal mood; 19.1% a single negative mood (3.3% depression, 10.4% anxiety, 5.4% stress);

Table 7.2 Percentage of co-occurrence of emotional states resulting by DASS scores

	Depressed		Not depressed	
	Stressed	Not stressed	Stressed	Not stressed
Anxious	17.1	4.2	10.8	10.4
Not anxious	4.2	3.3	5.4	44.6

19.2% two negative moods (4.2% depression and anxiety; 4.2% depression and stress; 10.8% stress and anxiety); 17.1% all three negative moods (depression, anxiety and stress). Scoring details can be found at the following link https://sites.google.com/site/becogsys/emothaw.

7.4 Database Evaluation

7.4.1 Evaluation Setup

A set of students were selected from the database described in Sect. 7.3 and divided in three dichotomous classes according to the DASS-42 scores:

D (Depression) Scale: depressed/not depressed group (DG/N-DG);
A (Anxiety) Scale: anxious/not anxious group (AG/N-AG);
S (Stress) Scale: stressed/not stressed group (SG/N-SG).

The three groups had the following characteristics:

D Scale: 42 subjects ($23m$ and $19f$) with depressed mood (DG) aged from 19 to 30 years ($M = 24.4$; $SD = 2.6$) and 42 subjects ($23m$ and $19f$) with non-depressed mood (N-DG) aged from 19 to 32 years ($M = 25.10$; $SD = 2.4$);

A Scale: 71 subjects ($33m$ and $38f$) with anxious mood (AG) aged from 19 to 30 years ($M = 24.30$; $SD = 2.4$) and 71 subjects ($33m$ and $38f$) with non-anxious mood (N-AG) aged from 18 to 32 years ($M = 24.69$; $SD = 2.5$);

S Scale: 50 subjects ($27m$ and $23f$) with stressed mood (SG) aged from 19 to 31 years ($M = 23.88$; $SD = 2.4$) and 50 subjects ($27m$ and $23f$) with non-stressed mood (N-SG) aged from 20 to 29 years ($M = 24.76$; $SD = 2.05$).

Subjects with negative mood (DG, AG and SG) had moderate, severe and extremely severe scores, while typical subjects (N-DG, N-AG, N-SG) had normal scores into the three (depression, anxious, stress) DASS scales (see Table 7.1).

In order to assess whether writing and drawing features (extracted from tasks 1, 2, 3, 6 and 7, described in Sect. 7.2.1), of subjects with negative (depressed, anxious and stressed) mood significantly differ from typical ones, two Mixed ANOVAs were performed. The first concerns timing-based and the second ductus-based features. Timing-based futures measures the time spent by the subject for each specific task on each pen state (Down, Up, Idle). Ductus-based features measure the number of times the subject has changed the pen state, while performing a task.

With respect to timing features, three $2 \times 3 \times 5$ mixed ANOVAs (one for each Scale: depression, anxiety, stress) were performed, with Group (depressed/not depressed (DG/N-DG); anxious/not anxious (AG/N-AG); stressed/not stressed (SG/N-SG)) as between factor, and features (tUp, tDown, tIdle) and tasks (I [pentagon drawing], II [house drawing], III [words in capital letter], VI [clock drawing], VII [writing of sentence]) as within factors.

With respect to the ductus features, three $2 \times 3 \times 5$ mixed ANOVAs (one for each Scale: depression, anxiety, stress) were performed, with Group (depressed/not depressed (DG/N-DG); anxious/not anxious (AG/N-AG); stressed/not stressed (SG/N-SG)) as between factor, and features (nUp, nDown, nIdle) and tasks (I [pentagon drawing], II [house drawing], III [words in capital letter], VI [clock drawing], VII [writing of sentence]) as within factors.

7.4.2 Results

Table 7.3 shows that 74% of depressed subjects is also anxious and stressed, 45% of anxious subjects is also depressed and stressed, and 80% of stressed subjects is also depressed and anxious, suggesting co-occurrences of the three negative moods. These co-occurrences seem contradict the aims of the DASS scale, whose "task was to develop an anxiety scale that would provide maximum discrimination from the BDI and other measures of depression" (Lovibond and Lovibond 1995 [13], p. 336). Further investigations are in order to assess whether these co-occurrences are indicating a real co-occurrence of the negative moods or overlapping factors among the three DASS scales defined into the DASS questionnaire.

7.4.2.1 Timing-Based Features

Table 7.4 reports on the rows the three DASS scales (depression, anxiety and stress), each partitioned in the associate negative (DG, AG, and SG) and normal (N-DG, N-AG, N-SG) mood respectively, and in turn each partitioned into the three pen-state timing features (tUp, tDown, tIdle). Columns report for these features the mean (M) and standard deviation (SD) for each of the 5 performed exercises. Data in bold indicates features where statistically significant differences are observed.

Table 7.3 Percentage of co-occurrence of emotional state in DG, AG, SG Group

DG	Anxious	Not anxious
Stressed	73.8	4.8
Not stressed	9.5	11.9
AG	Depressed	Not depressed
Stressed	45.1	28.2
Not stressed	5.6	21.1
SG	Depressed	Not depressed
Anxious	80.0	4.0
Not anxious	8.0	8.0

Table 7.4 Mean (M) and Standard Deviation (SD) of timing-based on features of each exercise. Data in bold indicates features that significantly differ ($p < .05$) in the binary categorization of each DASS scale, i.e., depressed/not depressed, anxious/not anxious, stressed/not stressed

DASS-42 scales	Groups	Features	Ex. I M	sd	Ex. II M	sd	Ex. III M	sd	Ex. VI M	sd	Ex. VII M	sd
Depression scale	DG	1(tUP)	8.39	1.12	14.66	1.13	14.09	0.70	13.86	1.05	9.83	0.70
	N-DG		6.16		12.95		12.65		13.30		9.31	
	DS	2(tDown)	**11.62**	0.79	19.06	1.18	**16.22**	0.49	13.50	0.71	**16.12**	0.43
	N-DG		**9.00**		16.87		**14.72**		11.80		14.95	
	DS	3(tidle)	**2.61**	0.41	4.55	0.48	3.44	0.32	5.41	0.80	3.17	0.33
	N-DG		**1.42**		3.37		3.03		5.90		2.93	
Anxiety scale	AG	1(tUP)	**8.47**	0.65	14.65	0.77	**14.09**	0.49	15.38	1.01	8.89	0.52
	N-AG		**6.25**		12.91		**12.70**		13.56		9.54	
	AG	2(tDown)	**11.58**	0.53	**19.27**	0.80	**16.25**	0.37	14.80	0.68	15.77	0.34
	N-AG		**9.03**		**15.57**		**14.58**		11.63		15.09	
	AG	3(tidle)	**2.56**	0.30	**5.04**	0.43	3.84	0.39	7.87	1.24	3.02	0.82
	N-AG		**1.46**		**3.10**		3.57		5.95		4.33	
Stress scale	SG	1(tUP)	6.94	0.70	13.46	0.76	13.45	0.58	13.84	0.78	9.14	0.77
	N-SG		5.72		12.71		12.33		13.12		9.60	
	SG	2(tDown)	10.31	0.55	17.89	0.90	15.21	0.38	13.07	0.58	15.31	0.41
	N-SG		9.06		15.28		14.52		11.47		15.18	
	SG	3(tidle)	2.16	0.31	**4.57**	0.39	3.73	0.38	5.78	0.64	3.38	1.16
	N-SG		1.28		**2.67**		3.51		5.23		4.96	

D Scale. A mixed ANOVA shows there are significant differences for Group [F(1,82)=4.26; p=.042; M(DG)=10.43 s; M(N-DG)=9.22; SD=0.41; MD=1.21 s]. The Group × Features × Exercise interaction shows significant differences:

- In the exercise I for features:

  - 2 (tDown) [F(1,82)=5.51; p=.021; M(DG)=11.61 s; M(N-DG)=8.99 s; SD=0.78; MD=2.62];
  - 3 (tIdle) [F(1,82)=4.17; p=.044; M(DG)=2.61 s; M(N-DG)=1.42 s; SD=0.41; MD=1.18];

- In the exercise III for features:

  - 2 (tDown) [F(1,82)=4.78; p=.032; M(DG)=16.22 s; M(N-DG)=14.71 s; SD=.47; MD=1.50].

A Scale. A mixed ANOVA shows significant differences for Group [F(1,140)= 11.31; p=.001; M(AG)=10.75 s; M(N-AG)=9.28; SD=0.31; MD=1.48 s]. The Group × Features × Exercise interaction shows significant differences:

- In the exercise I for features:

  - 1 (tUp) [F(1,140)=5.77; p=.018; M(AG)=8.46 s; M(N-AG)=6.25 s; SD=0.65; MD=2.21];
  - 2 (tDown) [F(1,140)=11.44; p=.001; M(AG)=11.58 s; M(N-AG)=9.02 s; SD= 0.53; MD=2.55];
  - 3 (tIdle) [F(1,140)=6.50; p=.012; M(AG)=2.55 s; M(N-AG)=1.45 s; SD=0.43; MD=1.09];

- In the exercise II for features:

  - 2 (tDown) [F(1,140)=10.80; p=.001; M(AG)=19.27 s; M(N-AG)=15.56 s; SD=0.79; MD=3.70];
  - 3 (tIdle) [F(1,140)=10.15; p=.002; M(AG)=5.04 s; M(N-AG)=3.09 s; SD=0.43; MD=1.94];

- In the exercise III for features:

  - 1 (tUp) [F(1,140)=4.03; p=.046; M(AG)=14.08 s; M(N-AG)=12.69 s; SD=0.49; MD=1.39];
  - 2 (tDown) [F(1,140)=10.41; p=.002; M(AG)=16.25 s; M(N-AG)=14.55 s; SD=0.36; MD=1.67];

- In the exercise VI for features:

  - 2 (tDown) [F(1,140)=10.82; p=.001; M(AG)=14.80 s; M(N-AG)=11.63 s; SD=0.68; MD=3.17].

S Scale. A Mixed ANOVA shows no significant differences for Group [F(1,98)= 3.55; p=.06; M(SG)=9.88 s; M(N-SG)=9.10; SD=0.29; MD=0.78 s]. The Group × Features × Exercise interaction shows significant differences:

- In the exercise II for features:

  - 2 (tDown) [F(1,98)=4.28; p=.041; M(AG)=17.89 s; M(N-AG)=15.26 s; SD=0.90; MD=2.63]
  - 3 (tIdle) [F(1,98)=11.44; p=.001; M(SG)=4.57 s; M(N-SG)=2.68 s; SD=0.39; MD=1.89];

7.4.2.2 Ductus-Based Features

Table 7.5 reports on the rows the three DASS scales (depression, anxiety and stress), each partitioned in the associate negative (DG, AG, and SG) and normal (N-DG, N-AG, N-SG) mood respectively, and in turn each partitioned into the three pen-state ductus features (nUp, nDown, nIdle). Columns report for these features the mean (M) and standard deviation (SD) for each of the 5 performed exercises. Data in bold indicates Features where statistically significant differences are observed.

D Scale. A mixed ANOVA shows significant differences for Group [F(1,82)=4.71; p=.033; M(DG)=28.09; M(N-DG)=25.40; SD=0.87; MD=2.70]. No significant differences are observed for Group × Feature × Exercise and Group × Exercise interactions. Conversely, Group × Feature interaction shows significant differences between DS and N-DS for the feature 2 (nDown) [F(1,82)=4.89; p=.030; M(DG)=35.07; M(N-DG)=31.76; SD=1.05; MD=3.03].

A Scale. A mixed ANOVA shows significant differences for Group $[F(1, 140) = 6.16; p = .014; M(AG) = 27.59; M(N - AG) = 25.34; SD = 0.63; MD = 2.22]$. The Group × Features × Exercise interaction shows significant differences:

- In the exercise I for features:

  - 1 (nUp) [F(1,140)=6.06; p=.015; M(AG)=12.16 s; M(N-AG)=7.69 s; SD=1.28; MD=4.48];

- In the exercise II for features:

  - 3 (nIdle) [F(1,140)=5.75; p=.018; M(AG)=20.73 sec; M(N-AG)=15.90 s; SD=1.42; MD=4.83].

S Scale. A mixed ANOVA shows significant differences for Group [F1,98=4.16; p=.044; M(SG)=26.88; M(N-SG)=24.95; SD=0.66; MD=1.92]. The Group × Features × Exercise interaction shows significant differences:

- In the exercise II for features:

  - 3 (nIdle) [F(1,98)=7.51; p=.007; M(SG)=21.28 s; M(N-SG)=15.12 s; SD=1.42; MD=6.16].

Table 7.5 Mean (M) and Standard Deviation (SD) of ductus-based on features of each exercise. Data in bold indicates features that significantly differ ($p < .05$) in the binary categorization of each DASS scale, i.e., depressed/not depressed, anxious/not anxious, stressed/not stressed

DASS-42 scales	Groups	Features	Ex. I		Ex. II		Ex. III		Ex. VI		Ex. VII	
			M	sd	M	sd	M	sd	M	sd	M	sd
Depression scale	DS	1(tUP)	12.93	2.03	27.05	1.84	62.31	1.23	25.74	1.05	45.29	1.80
	N-DS		7.40		23.69		61.07		24.62		42.02	
	DS	2(tDown)	**13.29**	**2.00**	**27.95**	**1.78**	**62.69**	**1.39**	**26.40**	**1.05**	**45.93**	
	N-DS		**7.90**		**24.26**		**60.05**		**25.12**		**41.50**	
	DS	3(tidle)	10.64	1.60	21.40	2.22	25.74	1.14	19.86	2.11	10.05	0.93
	N-DS		8.12		16.90		9.17		18.02		11.14	
Anxiety scale	AS	1(tUP)	12.17	1.28	26.14	1.06	60.75	0.70	27.97	1.51	41.69	1.17
	N-AS		7.69		23.44		60.39		24.85		42.37	
	AS	2(tDown)	12.59	1.27	26.38	1.06	60.90	0.82	28.54	1.49	41.66	1.18
	N-AS		8.21		23.96		58.87		25.35		41.59	
	AS	3(tidle)	11.23	1.11	**20.73**	1.42	10.44	0.82	22.46	2.12	10.27	0.80
	N-AS		8.48		**15.90**		10.06		17.87		11.51	
Stress scale	SG	1(tUP)	10.62	1.64	24.80	1.25	61.28	0.99	25.70	1.02	42.96	1.62
	N-SG		7.12		22.76		59.86		25.24		42.22	
	SG	2(tDown)	11.18	1.62	25.00	1.25	61.72	1.08	26.30	1.02	42.96	1.61
	N-SG		7.64		23.26		58.82		25.72		41.94	
	SG	3(tidle)	8.80	1.14	**21.28**	1.29	10.54	1.00	19.12	1.65	10.40	0.90
	N-SG		7.14		**15.12**		9.70		16.62		11.20	

7.5 Conclusion

This paper present an enlarged and refined handwriting database for detecting negative moods (depression, anxiety and stress), and introduces the extraction of a novel time-based handwriting feature indicating the idle subject state. Mixed ANOVA analyses, based on a binary categorization of the groups, reveal significant differences among features collected from subjects with negative moods with respect to the control group depending on the involved exercises and feature categories (in time or frequency of the considered events). In addition, our preliminary investigation shows that the type of negative mood (depressed, anxious or stressed) affects the drawing and handwriting tasks in a different way, considering that some exercises and features are important to detect one specific negative mood but not the others. As a future development, a plan to exploit similar features in order to detect personality traits is featured. Can some personality traits be extracted from handwriting?

Acknowledgements The research leading to the results presented in this paper has been conducted in the project EMPATHIC that received funding from the European Unions Horizon 2020 research and innovation programme under grant agreement number 769872.

References

1. Antony, M.M., Bieling, P.J., Cox, B.J., Enns, M.W., Swinson, R.P.: Psychometric properties of the 42-item and 21-item versions of the depression anxiety stress scales in clinical groups and a community sample. Psychol. Assess. **10**(2), 176–181 (1998)
2. Brown, T.A., Chorpita, B.F., Korotitsch, W., Barlow, D.H.: Psychometric properties of the depression anxiety stress scales (DASS) in clinical samples. Behav. Res. Therapy **35**(1), 79–89 (1997)
3. Chen, C.-C., Aggarwal, J.K.: Modeling human activities as speech. CVPR **2011**, 3425–3432 (2011)
4. Clara, I.P., Cox, B.J., Enns, M.W.: Confirmatory factor analysis of the depression-anxiety-stress scales in depressed and anxious patients. J. Psychopathol. Behav. Assess. **23**(1), 61–67 (2001)
5. Cordasco, G., Esposito, M., Masucci, F., Riviello, M.T., Esposito, A., Chollet, G., Schlgl, S., Milhorat, P., Pelosi. G.: Assessing voice user interfaces: the vassist system prototype. In: 2014 5th IEEE Conference on Cognitive Infocommunications (CogInfoCom), pp. 91–96 (2014)
6. Cordasco, G., Riviello, M.T., Capuano, V., Baldassarre, I., Esposito, A.: Youtube emotional database: how to acquire user feedback to build a database of emotional video stimuli. In: 2013 IEEE 4th International Conference on Cognitive Infocommunications (CogInfoCom), pp. 381–386 (2013)
7. Crawford, J.R., Henry, J.D.: The depression anxiety stress scales (DASS): normative data and latent structure in a large non-clinical sample. Br. J. Clin. Psychol. **42**(2), 111–131 (2003)
8. Esposito, A., Esposito, A.M., Vogel, C.: Needs and challenges in human computer interaction for processing social emotional information. Pattern Recogn. Lett. **66**(C), 41–51 (2015)
9. Folstein, M.F., Folstein, S.E., McHugh, P.R.: "Mini-mental state". A practical method for grading the cognitive state of patients for the clinician. J. Psychiatr. Res. **12**(3), 189–198 (1975)

10. Hyungsin, K.: The clockme system: computer-assisted screening tool for dementia. Ph.D. Dissertation, College Computing, Georgia Institute of Technology (2013)
11. Kline, P.: The Handbook of Psychological Testing. Routledge, London, New York (2000)
12. Likforman-Sulem, L., Esposito, A., Faundez-Zanuy, M., Clmenon, S., Cordasco, G.: EMOTHAW: a novel database for emotional state recognition from handwriting and drawing. IEEE Trans. Hum.-Mach. Syst. **47**, 273–284 (2016)
13. Lovibond, P.F., Lovibond, S.H.: The structure of negative emotional states: comparison of the depression anxiety stress scales (DASS) with the beck depression and anxiety inventories. Behav. Res. Therapy **33**(3), 335–343 (1995)
14. Luria, G., Kahana, A., Rosenblum, S.: Detection of deception via handwriting behaviors using a computerized tool: toward an evaluation of malingering. Cogn. Comput. **6**(4), 849–855 (2014)
15. Maldonato, M., Dell'Orco, S., Esposito, A.: The emergence of creativity. World Futures **72**(7–8), 319–326 (2016)
16. Neils-Strunjas, J., Shuren, J., Roeltgen, D., Brown, C.: Perseverative writing errors in a patient with alzheimer's disease. Brain Lang. **63**(3), 303–320 (1998)
17. O'Reilly, C., Plamondon, R.: Design of a neuromuscular disorders diagnostic system using human movement analysis. In: 11th International Conference on Information Science, Signal Processing and their Applications, ISSPA 2012, Montreal, QC, Canada, 2–5 July 2012, pp. 787–792 (2012)
18. Plamondon, R., Srihari, S.N.: On-line and off-line handwriting recognition: a comprehensive survey. IEEE Trans. Pattern Anal. Mach. Intell. **22**(1), 63–84 (2000)
19. Plamondon, R., O'Reilly, C., Ouellet-Plamondon, C.: Strokes against stroke–strokes for strides. Pattern Recogn. **47**(3), 929–944 (2014)
20. Riviello, M.T., Capuano, V., Ombrato, G., Baldassarre, I., Cordasco, G., Esposito, A.: In: The Influence of Positive and Negative Emotions on Physiological Responses and Memory Task Scores, pp. 315–323. Springer, Bassis, Simone and Esposito, Anna and Morabito, Francesco Carlo, Cham (2014)
21. Rosenblum, S., Parush, S., Weiss, P.L.: The in air phenomenon: temporal and spatial correlates of the handwriting process. Percept. Mot. Skills **96**, 933–9545 (2003)
22. Scibelli, F., Troncone, A., Likforman-Sulem, L., Vinciarelli, A., Esposito, A.: How major depressive disorder affects the ability to decode multimodal dynamic emotional stimuli. Front. in ICT **3**, 16 (2016)
23. Sesa-Nogueras, E., Faundez-Zanuy, M., Mekyska, J.: An information analysis of in-air and on-surface trajectories in online handwriting. Cogn. Comput. **4**(2), 195–205 (2012)
24. Severino, G.A., Haynes, W.D.G.: Development of an italian version of the depression anxiety stress scales. Psychol. Health Med. **15**(5), 607–621 (2010). PMID: 20835970
25. Tang, T.L.-P.: Detecting honest people's lies in handwriting. J. Bus. Ethics **106**(4), 389–400 (2012)
26. Troncone, A., Palumbo, D., Esposito. A.: Mood Effects on the Decoding of Emotional Voices. pp. 325–332. Springer, Bassis, Simone and Esposito, Anna and Morabito, Francesco Carlo, Cham (2014)

Chapter 8
Content-Based Music Agglomeration by Sparse Modeling and Convolved Independent Component Analysis

Mario Iannicelli, Davide Nardone, Angelo Ciaramella and Antonino Staiano

Abstract Music has an extraordinary ability to evoke emotions. Nowadays, the music fruition mechanism is evolving, focusing on the music content. In this work, a novel approach for agglomerating songs on the basis of their emotional contents, is introduced. The main emotional features are extracted after a pre-processing phase where both Sparse Modeling and Independent Component Analysis based methodologies are applied. The approach makes it possible to summarize the main sub-tracks of an acoustic music song (e.g., information compression and filtering) and to extract the main features from these parts (e.g., music instrumental features). Experiments are presented to validate the proposed approach on collections of real songs.

8.1 Introduction

Nowadays, one of the main channels for accessing information about people and their social interactions is multimedia content (pictures, music, videos, e-mails, etc.) [16]. Emotions have a fundamental role in rational decision-making, perception, human interaction, human intelligence [12], and the principal aspect of music is to evoke emotions [1, 2]. In literature, emotion consists of a short duration (seconds to minutes), while mood has a longer duration (hours or days), and the issue of recognizing their features in music tracks is challenging. In [12], an hierarchical framework for mood detection from acoustic music data by following some music psychological

M. Iannicelli · D. Nardone · A. Ciaramella (✉) · A. Staiano
Dipartimento di Scienze e Tecnologie, Università degli Studi di Napoli "Parthenope",
Isola C4, Centro Direzionale, I-80143 Napoli (NA), Italy
e-mail: angelo.ciaramella@uniparthenope.it

M. Iannicelli
e-mail: mario.iannicelli@studenti.uniparthenope.it

D. Nardone
e-mail: davide.nardone@studenti.uniparthenope.it

A. Staiano
e-mail: antonino.staiano@uniparthenope.it

© Springer International Publishing AG, part of Springer Nature 2019
A. Esposito et al. (eds.), *Quantifying and Processing Biomedical and Behavioral Signals*, Smart Innovation, Systems and Technologies 103,
https://doi.org/10.1007/978-3-319-95095-2_8

theories in western cultures, is presented. In [4, 14, 17] the authors propose fuzzy models to determine emotion or mood classes. Recently, several community web-sites that combine social interactions with music and entertainment exploration have been proposed. For example, *Stereomood* [15] is a free emotional internet radio that suggests the music that best suits mood and daily activities of an user. It allows the users to create play-lists for different occasions and to share emotions through a manual tagging process.

In this work we propose a system for songs agglomeration from the extracted emotional contents. The extraction of the features is accomplished by a pre-processing phase, where both Sparse Modeling (SM) and Independent Component Analysis (ICA) based methodologies are applied. SM permits to select the representative sub-tracks of an acoustic music song obtaining information compression and filtering. ICA allows estimating the fundamental components from these sub-parts, extracting the main sub-tracks features (e.g., correlated to music instrumental).

The paper is organized as follows. In Sect. 8.2, music emotional features are described. Sections 8.3 and 8.4, introduce the Sparse Modeling and Convolved Independent Component Analysis methodologies, respectively. In Sect. 8.5, a description of the overall system is given. Section 8.6, describes several experimental tasks also illustrating their results. Finally, in Sect. 8.7, a concluding discussion closes the paper.

8.2 Emotional Music Content Characterization

In order to properly characterize emotions from acoustic music signals, one can consider fundamental indices such as *intensity*, *rhythm*, *key*, *harmony* and *spectral centroid* [4, 14, 17]. In the following, a deepen description is provided for each index.

8.2.1 Intensity

The intensity of sound sensation is related to the amplitude of sound waves [9]. In general, sadness is associated to low intensity, such as melancholy, tenderness or peacefulness. High intensity is associated to positive emotions like joy, excitement or triumph. Very high intensity with many variations during the time could be associated with anger or fear.

8.2.2 Rhythm

The rhythm of a song is evaluated by analysing the beat and tempo [6]. The beat is a regularly occurring pattern of rhythmic stresses in music, and tempo is the beat speed, usually expressed in Beats Per Minute (BPM). Fast music causes blood pressure, heart and breathing rate to go up, while slow music causes these to drop. Moreover, regular beats make listeners peaceful or even melancholic, but irregular beats could make some listeners feel aggressive or unsteady.

8.2.3 Key

A scale is a group of pitches (scale degrees) arranged in ascending order. These pitches span an octave and scale patterns can be duplicated at any pitch. Rewriting the same scale pattern at a different pitch is named transposition. A key is the major or minor scale around which a piece of music revolves. In order to characterize the scale in our system, the Key Detection algorithm proposed in [13] is used. The algorithm returns the estimated key for each key change. We consider the key of the song as the key associated with the maximum duration in the song.

8.2.4 Harmony and Spectral Centroid

Harmonics can be observed perceptually when harmonic musical instruments are performed in a song. Harmony refers to the way chords are constructed and how they follow each other. Since this analysis of the harmony does not consider the fundamental pitch of the signal, we also consider the *spectral centroid* [10].

8.3 Representative Subtracks

In the proposed system, we consider a Sparse Modeling (SM) for extracting significative parts of a music song. In particular, in a SM a data matrix $\mathbf{Y} = [\mathbf{y}_1, \ldots, \mathbf{y}_N]$, where $\mathbf{y}_i \in R^m, i = 1, \ldots, N$, is considered [8]. The main objective is to find a compact dictionary $\mathbf{D} = [\mathbf{d}_1, \ldots, \mathbf{d}_N] \in R^{m \times N}$ and coefficients $\mathbf{X} = [\mathbf{x}_1, \ldots, \mathbf{x}_N] \in R^{N \times N}$, for efficiently representing the collection of data points $\mathbf{Y}$. The best representation of the data is obtained by minimizing the following objective function

$$\sum_{i=1}^{N} \|\mathbf{y}_i - \mathbf{D}\mathbf{x}_i\|_2^2 = \|\mathbf{Y} - \mathbf{D}\mathbf{X}\|_F^2, \tag{8.1}$$

with respect to the dictionary $\mathbf{D}$ and the coefficient matrix $\mathbf{X}$, subject to appropriate constraints. In the sparse dictionary learning framework, one requires the coefficient matrix $\mathbf{X}$ to be sparse by solving

$$\min_{D,X} \|\mathbf{Y} - \mathbf{DX}\|_F^2$$

$$\text{s.t. } \|\mathbf{x}_i\|_0 \leq s, \|\mathbf{d}_j\|_2 \leq 1 \quad \forall i, j, \tag{8.2}$$

where $\|\mathbf{x}_i\|_0$ indicates the number of nonzero elements of $\mathbf{x}_i$. In particular, dictionary and coefficients are learned simultaneously as such that each data point $\mathbf{y}_i$ is written as a linear combination of at most s atoms of the dictionary [3]. Now, it can be noticed that the dictionary learning framework can be evaluated such that representative points coincide with some of the actual data points. The reconstruction error of each data point can be expressed as a linear combination of all data

$$\sum_{i=1}^{N} \|\mathbf{y}_i - \mathbf{Yc}_i\|_2^2 = \|\mathbf{Y} - \mathbf{Y}C\|_F^2, \tag{8.3}$$

with respect to the coefficient matrix $\mathbf{C} \triangleq [\mathbf{c}_1, \ldots, \mathbf{c}_N] \in R^{N \times N}$. To find $k \ll N$ representatives we use the following optimization problem

$$\min \|\mathbf{Y} - \mathbf{Y}C\|_F^2$$

$$\text{s.t. } \|\mathbf{C}\|_{0,q} \leq k, \mathbf{1}^{\mathbf{T}}\mathbf{C} = \mathbf{1}^{\mathbf{T}}, \tag{8.4}$$

where $\|\mathbf{C}\|_{0,q} \triangleq \sum_{i=1}^{N} I(\|\mathbf{C}\|_q) > 0$, $\mathbf{c}^i$ denotes the i-th row of $\mathbf{C}$ and $I(.)$ denotes the indicator function. In particular, $\|\mathbf{C}\|_{0,q}$ counts the number of nonzero rows of C. Since this is an NP-hard problem, a standard l_1 relaxation of this optimization is adopted

$$\min \|\mathbf{Y} - \mathbf{Y}C\|_F^2$$

$$\text{s.t. } \|\mathbf{C}\|_{1,q} \leq \tau, \mathbf{1}^{\mathbf{T}}\mathbf{C} = \mathbf{1}^{\mathbf{T}}, \tag{8.5}$$

where $\|\mathbf{C}\|_{1,q} \triangleq \sum_{i=1}^{N} \|\mathbf{c}^i\|_q$ is the sum of the l_q norms of the rows of C, and $\tau > 0$ is an appropriately chosen parameter. The solution of the optimization problem 8.5, not only indicates the representatives as the nonzero rows of C, but also provides information about the ranking, i.e., relative importance of the representatives for describing the dataset. We can rank k representatives $\mathbf{y}_{i_1}, \ldots, \mathbf{y}_{i_k}$ as $i_1 \geq i_2 \geq \cdots \geq i_k$, i.e., $\mathbf{y}_{i_1}$ has the highest rank and $\mathbf{y}_{i_k}$ has the lowest rank. In this work, by using the Lagrange multipliers, the optimization problem is defined as

$$\min \frac{1}{2}\|\mathbf{Y} - \mathbf{Y}C\|_F^2 + \lambda\|\mathbf{C}\|_{1,q}$$

$$\text{s.t. } \mathbf{1}^{\mathbf{T}}\mathbf{C} = \mathbf{1}^{\mathbf{T}}. \tag{8.6}$$

The algorithm is implemented by using an Alternating Direction Method of Multipliers (ADMM) optimization framework (see [8] for further details).

8.4 Blind Source Separation and ICA

In signal processing, Independent Component Analysis (ICA) is a computational method for separating a multivariate signal into additive components, particularly adopted for Blind Source Separation of instantaneous mixtures [11]. In various real-world applications, convolved and time-delayed versions of the same sources can be observed instead of instantaneous ones [5, 11]. This is due to multipath propagation, typically caused by reverberations from obstacles. To model this scenario, a convolutive mixture model must be considered. In particular, each element of the mixing matrix $\mathbf{A}$ in the model $\mathbf{x}(t) = \mathbf{As}(t)$, is a filter rather than a scalar, as in the following equation

$$x_i(t) = \sum_{j=1}^{n} \sum_{k} a_{ikj} s_j(t - k), \qquad (8.7)$$

for $i = 1, \ldots, n$. To invert the convolutive mixtures $x_i(t)$, a set of similar FIR filters should be used

$$y_i(t) = \sum_{j=1}^{n} \sum_{k} w_{ikj} x_j(t - k). \qquad (8.8)$$

The output signals $y_1(t), \ldots, y_n(t)$ of the separating system are the estimates of the source signals $s_1(t), \ldots, s_n(t)$ at discrete time t, and w_{ikj} are the coefficients of the FIR filters of the separating system. In this paper, in order to estimate the w_{ikj} coefficients we adopt the approach introduced in [5] (named Convolved ICA, CICA). The main idea is to use a Short Time Fourier Transform (STFT) for moving the convolved mixtures in the frequency domain. In particular, the observed mixtures are divided in frames (which usually overlap each other, to reduce artifacts at the boundary) to obtain $(X_i(\omega, t))$, both in time and frequency. For each frequency bin, we get n observations to which apply the ICA models in the complex domain. To solve the permutation indeterminacy [11], an Assignment Problem (e.g., Hungarian algorithm) with a Kullback-Leibler divergence is adopted [5].

8.5 System Architecture

As earlier stated, the aim is to agglomerate songs by considering their emotional contents. To accomplish this task a pre-processing step must be performed. A prototype of the proposed pre-processing system is described in Fig. 8.1. In detail, a music track is divided in several frames to obtain a matrix $\mathbf{Y}$ of observations. The matrix is used in a SM approach, as described in Sect. 8.3 (i.e., Eq. 8.6), to obtain the representative frames (sub-tracks) of the overall song track. This summarizing step is useful for improving information storage (e.g., for mobile) and to avoid unnecessary information. Successively, the CICA approach of Sect. 8.4 is adopted to separate the compo-

Fig. 8.1 Pre-processing procedure of the proposed system

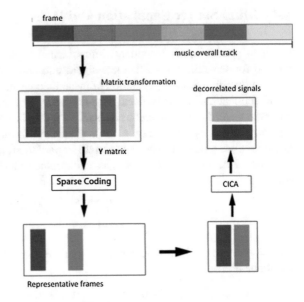

nents from the representative extracted sub-tracks. CICA permits to extract the independent components characterizing the intrinsic information of the songs, for example those related to singer voice and music instrumentals. Successively, for each computed component the emotional features, described in Sect. 8.2, are extracted. Finally, an Hierarchical clustering is used to agglomerate the extracted information [7].

8.6 Experimental Results

Now we report some experimental results obtained by using the music emotion recognition system described in Sect. 8.5 on two different datasets. We tested our system by considering the first 120 seconds of the songs with a sampling frequency of 44,100 Hz and 16 bit of quantization. The agglomeration results of three criteria are compared, namely

1. overall song elaboration;
2. applying SM;
3. applying SM and CICA.

In the first experiment, we consider a dataset composed by 9 popular songs as listed in Table 8.1. For agglomeration purposes, we applied an hierarchical clustering with Euclidean distance and a complete linkage. In Fig. 8.2, the agglomeration obtained by using the three different criteria are shown (Fig. 8.2a–c), respectively. Comparing the results, we note that, in all cases, songs with labels 1, 9 and 6 get agglomerated together for their well defined musical content (e.g., rhythm). The main agglom-

Table 8.1 Songs used for the first experiment

Author	Title	Label
AC/DC	Back in Black	1
Nek	Almeno stavolta	2
Led Zeppelin	Stairway to Heaven	3
Louis Armstrong	What a wonderful world	4
Madonna	Like a Virgin	5
Michael Jackson	Billie Jean	6
Queen	The Show Must Go On	7
The Animals	The House of the Rising Sun	8
Sum 41	Still Waiting	9

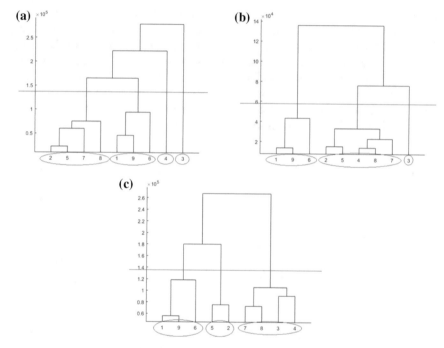

Fig. 8.2 Hierarchical clustering on the dataset of 9 songs applying three criteria: **a** overall song elaboration; **b** sparse modeling; **c** sparse modeling and CICA

eration differences are highlighted when the musical instruments contained in the songs, are considered. As an example, songs 3 (without its last part) and 4, by using SM and CICA (Fig. 8.2c), get clustered together due to the rhythmic content and the presence in 3 of a predominant synthesized wind musical instrument, as the wind musical instruments in song 4. Moreover, this cluster is close to the cluster composed by songs 7 and 8 since they share a musical keyboard content.

In the second experiment, we consider a dataset composed by 28 songs of different genres, namely

- 10 *children songs*,
- 10 *classical music*,
- 8 *easy listening* (multi-genre class).

In Fig. 8.3, we show the agglomeration obtained by using the three criteria previously described. In this experiment, comparing the results, a first consideration is about song number 4. Analyzing Fig. 8.3a (overall song elaboration) and Fig. 8.3b (sparse modeling) we deduce that the song number 4 is in a different agglomerated cluster. Analyzing the songs in the clusters we observe that in the first case we obtain a wrong result. In particular, by observing the waveform of song 4 (see Fig. 8.4), that song exhibits two different loudnesses and the extraction of the emotional features on the overall song is not accurate. In this case, the SM extracts the representative frames obtaining a more robust estimation. Moreover, by applying CICA we also obtain the agglomeration of children and classic songs in two main classes (Fig. 8.3c). The first cluster gets separated in two subclasses, namely classic music and easy listening. In the second cluster, we find all children songs except songs 1 and 5. The misclassification of song 1 is due to the instrumental feature of the song (without a singer voice), like a classical music, whereas song 5 gets classified as easy listening because it is a children song with an adult man singer voice.

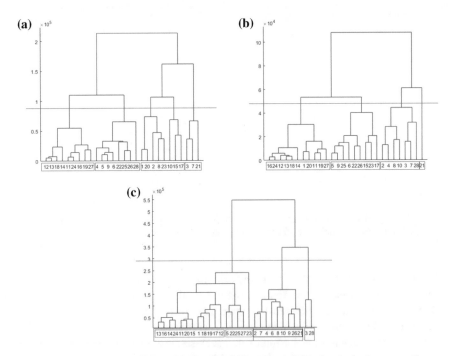

Fig. 8.3 Hierarchical clustering on the dataset of 28 songs applying three criteria: **a** overall song elaboration; **b** sparse modeling; **c** sparse modeling and CICA

Fig. 8.4 Waveform of song 4

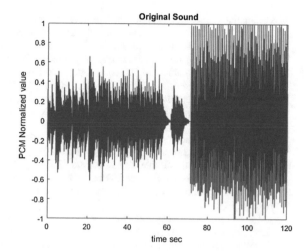

8.7   Conclusions

In this work, a system for music songs grouping from extracted emotional contents has been proposed. It is based on a pre-processing step consisting of a Sparse Modeling and Independent Component Analysis based approaches. This phase permits to select the representative subtracks of an acoustic music song and the estimation of the fundamental components from them. The experimental results show how the methodology allows obtaining a more precise content based agglomeration. In the future the authors will focus on the application of the approach on larger datasets and for classification tasks.

Acknowledgements The research was entirely developed when Mario Iannicelli was a Bachelor Degree student in Computer Science at University of Naples Parthenope. The authors would like to thank Marco Gianfico for his support and comments. This work was partially funded by the University of Naples Parthenope (*Sostegno alla ricerca individuale per il triennio 2016–2018* project).

References

1. J. L. Barrow-Moore, The Effects of Music Therapy on the Social Behavior of Children with Autism, Master of Arts in Education College of Education California State University San Marcos, November, 2007
2. Blood, A.J., Zatorre, R.J., Bermudez, P., Evans, A.C.: Emotional responses to pleasant and unpleasant music correlate with activity in paralimbic brain regions. Nature Neuroscience **2**, 382–387 (1999)
3. Ciaramella, A., Gianfico, M., Giunta, G.: Compressive sampling and adaptive dictionary learning for the packet loss recovery in audio multimedia streaming. Multimedia Tools and Applications **75**(24), 17375–17392 (2016)

4. Ciaramella, A., Vettigli, G.: Machine Learning and Soft Computing Methodologies for Music Emotion Recognition. Smart Innovation, Systems and Technologies **19**, 427–436 (2013)
5. A. Ciaramella, E. De Lauro, M. Falanga, S. Petrosino, Automatic detection of long-period events at Campi Flegrei Caldera (Italy), Geophysical Research Letters, 38 (18), 2013
6. Davies, M.E.P., Plumbley, M.D.: Context-dependent beat tracking of musical audio. IEEE Transactions on Audio, Speech and Language Processing. **15**(3), 1009–1020 (2007)
7. R. O. Duda, P. E. Hart, D. G. Stork, *Pattern Classification*, Wiley-Interscience, 2000
8. Elhamifar, E., Sapiro, G., Vidal, R. See all by looking at a few: Sparse modeling for finding representative objects (2012), Proceedings of the IEEE Computer Society Conference on Computer Vision and Pattern Recognition, art. no. 6247852, pp. 1600-1607
9. G. Revesz, Introduction to the psychology of music, Courier Dover Publications, 2001
10. Grey, J.M., Gordon, J.W.: Perceptual effects of spectral modifications on musical timbres. Journal of the Acoustical Society of America **63**(5), 1493–1500 (1978)
11. Hyvarinen, A., Karhunen, J., Oja, E.: Independent Component Analysis. John Wiley, Hoboken, N. J. (2001)
12. L. Lu, D. Liu, H.-J. Zhang, Automatic Mood Detection and Tracking of Music Audio Signals, IEEE Transaction on Audiom Speech, and Language Processing, vol. 14, no. 1, 2006
13. K. Noland and M. Sandler, Signal Processing Parameters for Tonality Estimation. In Proceedings of Audio Engineering Society 122nd Convention, Vienna, 2007
14. S. Jun, S. Rho, B.-J. Han, E. Hwang, A Fuzzy Inference-based Music Emotion Recognition System., Visual Information Engineering 2008 - VIE 2008, 5th International Conference on In Visual Information Engineering, 2008
15. *stereomood website*. http://www.stereomood.com
16. Vinciarelli, A., Pantic, M., Heylen, D., Pelachaud, C., Poggi, I., D'Errico, F.: Marc Schroeder. A Survey of Social Signal Processing, IEEE Transactions on Affective Computing, Bridging the Gap Between Social Animal and Unsocial Machine (2011)
17. Yang, Y.-H., Liu, C.-C., Chen, H.H.: Music Emotion Classification: A Fuzzy Approach. Proceedings of ACM Multimedia **2006**, 81–84 (2006)

Chapter 9
Oressinergic System: Network Between Sympathetic System and Exercise

Vincenzo Monda, Raffaele Sperandeo, Nelson Mauro Maldonato,
Enrico Moretto, Silvia Dell'Orco, Elena Gigante, Gennaro Iorio
and Giovanni Messina

Abstract Sport, in different ways, change considerably people's life. The purpose of this experiment was to reveal possible association between the stimulation of sympathetic system induced by exercise and the one induced by the rise of systemic concentration of Orexin A and bring the truth about orexins and sport network. Blood samples were collected from subjects (men, n $=$ 10; age: 23.2 ± 2.11 years) 15, 0 min before the start of exercise, and 30, 45, 60 min after a cycle ergometer exercise at 75 W for 15 min. Also, heart rate (HR), galvanic skin response (GSR), and rectal temperature were monitored. The exercise produce a significant rise ($p < 0.01$) in plasmatic orexin A with a peak at 30 min after the exercise bout, in association with a rise of the other three monitored variables: HR ($p < 0.01$), GSR ($p < 0.05$), and rectal temperature ($p < 0.01$). Our results indicate that plasmatic orexin A is involved in the reaction to physical activity and in the beneficial effects of sport.

Keywords Orexin · Physical exercise · Sport · Heart rate · Galvanic skin response · Rectal temperature · Sympathetic nervous system

V. Monda
Department of Experimental Medicine, Università degli Studi della Campania, Naples, Italy
e-mail: vincenzo.monda@unicampania.it

R. Sperandeo (✉) · E. Moretto · S. Dell'Orco · E. Gigante · G. Iorio
SiPGI Postgraduate School of Integrated Gestalt Psychotherapy, Naples, Italy
e-mail: raffaele.sperandeo@gmail.com

N. M. Maldonato · G. Messina
Department of Neuroscience and Reproductive and Odontostomatological
Sciences, University of Naples Federico II, Naples, Italy

V. Monda · R. Sperandeo · N. M. Maldonato · E. Moretto · S. Dell'Orco · E. Gigante
G. Iorio · G. Messina
Department of Clinical and Experimental Medicine, University of Foggia, Foggia, Italy

© Springer International Publishing AG, part of Springer Nature 2019
A. Esposito et al. (eds.), *Quantifying and Processing Biomedical and Behavioral Signals*, Smart Innovation, Systems and Technologies 103,
https://doi.org/10.1007/978-3-319-95095-2_9

9.1   Introduction

Orexins are synthesized in the lateral hypothalamic and perifornical areas [1, 2]. Orexin A and B are excitatory hypothalamic neuropeptides playing a relevant role in different physiologic functions, in fact Orexin neurons are called "multi-tasking" neurons [3, 4]. Despite recent studies have shown the role of the orexins in sleep and wakefulness and arousal system [5], thermoregulation, energetic homeostasis, control of energy metabolism [6], cardiovascular responses, feeding behavior [7], spontaneous physical activity (SPA), reward mechanisms, mood and emotional regulation and drug addiction [7–13], the function of orexins in metabolism pathways are far to be completely understood. Orexin A and Orexin B are neuropeptides composed respectively of 33 and 28 amino acids, the N-terminal portion presents more variability, whilst the C-terminal portion is similar between the two subtypes. The orexins activity is modulated by their specific receptors (OX1R, OX2R).

OX1R, higher affinity for orexin A than B, is distributed in PVT, anterior hypothalamus, prefrontal and infralimbic cortex (IL), stria terminalis bed nucleus (BST), hippocampus (CA2), amygdala, dorsal raphe (DR), ventral tegmental area (VTA), locus coeruleus (LC), and laterodorsal tegmental nucleus (LDT)/pedunculopontine nucleus (PPT) and transmits signals throughout the G-protein class activating a cascade that leads to an increase in intracellular calcium concentration [14, 15]; OX2R, similar affinities for the two subtypes, is distributed in amygdala, TMN, Arc, dorsomedial hypothalamic nucleus (DMH), LHA, BST, paraventricular nucleus (PVN), PVT, LDT/PPT, DR, VTA, CA3 in the hippocampus, and medial septal nucleus [14] and is probably associated to a G inhibitory protein class [16].

Orexin A could be also found in plasma, but its peripheral origin is not well-known. The endocrine pancreas seems to be the probable source of plasmatic orexin A, and the b-cells are retained secretor cells of this peptide [17]. Orexin A is strongly involved in the regulation of autonomic reactions so much that is possible to notice, after an intracerebroventricular (ICV) injection, tachycardia [18], associated with an increase metabolic rate [19] and blood pressure (BP) [20]. Furthermore, an ICV injection of Orexin A induces a rise in body temperature [21], and the contemporaneous hyperthermia and tachycardia suggests its widespread stimulation of the sympathetic nervous system. Microinjections into the nucleus of the solitary tract elicit dose dependent changes in HR and blood pressure. Stimulation of the hypothalamic perifornical region in orexin-ataxin mice with depleted orexin produced smaller and shorter lasting increases in HR and blood pressure than in control mice [22]. Exercise generates stimulation of sympathetic activity and temperature rise too. The purpose of this experiment was to reveal possible association between the stimulation of sympathetic system induced by exercise and the one induced by the rise of systemic concentration of orexin A and bring the truth about orexins and sport network.

9.2   Materials and Methods

9.2.1   Subjects

Ten healthy sedentary men (age: 23.2 ± 2.11 years) were recruited, in accordance with these criteria of inclusion, among those who contacted the Clinical Unit of Dietetics and Sports Medicine of the University of Campania "Luigi Vanvitelli". They were volunteers. None of the subjects was taking any medication and each subject was instructed to avoid beverages containing alcohol or caffeine. Their body weight was stable in the last year, none of the subjects was smoker or taking any medication and each subject was instructed to avoid strenuous physical activity and beverages containing alcohol or caffeine 7 days before the experimental procedure. Anthropometric values, expressed as mean $\pm$ standard deviation (SD), are reported in Table 9.1.

9.2.2   Ethics Statement

The experimental procedures followed the rule approved by Ethics Committee of the University of Campania "Luigi Vanvitelli". Patients were informed on the research and permission for the use of serum samples was obtained. All procedures conformed to the directives of the Declaration of Helsinki.

9.2.3   Study Protocol

The study protocol consisted of 1 day of testing in which each participant was asked to continue his normal daily activities and job. The subjects were required to consume food and beverages as usual (except drinks containing alcohol or caffeine) and to sleep enough hours. The period of experiment was divided into 3 times: resting time (0–15 min), exercise time (16–30 min), and recovery time (31–60 min). Five blood samples were carried out in all periods of experiment. Two samples at resting time (to demonstrate stable basal values), one sample at last minute of exercise, and two samples at recovery time. The physical activity consisted of a cycle ergometer exercise, performed at 16–30 min period. Exercise was same for all subjects: 75 W and 60 rpm for 15 min, with a relative intensity of 70%. All tests were performed in laboratory room with a normal ambient climate.

Table 9.1 Anthropometric values

Age (years)	BMI (kg/m^2)	BP systolic/diastolic (mmHg)
23.2 ± 2.11	21.01 ± 0.8	$115 \pm 5/68 \pm 7$

9.3 Measurement of Heart Rate, Galvanic Skin Response, and Rectal Temperature

The participants cycled on a calibrated mechanically braked cycle ergometer (Kettler ergometer E7, Fitness Service, CassanoMagnago, VA, Italy) at 75 W, nonstop for 15 min. The HR and galvanic skin response (GSR) data were recorded and annotated during all times of experiment. Each subject has been endowed with a chest belt hardwired to a digital R–R recorder (BTL08 SD ECG, BTL Industries, Varese (VA), Italy), where the QRS signal wave form R-R signal was sampled at the resolution of 1 ms. The HR (beats/min) was derived from the following formula: HR = 60 R-R interval-1; where, R-R interval was converted into seconds. The GSR parameters were simultaneously measured using the SenseWear Pro ArmbandTM (version 3.0, BodyMedia, Inc. PA, USA), which was worn on the right arm over the triceps muscle at the midpoint between acromion and olecranon processes, as recommended by the manufacturer. Rectal temperature was measured with electronic thermometer thermistor/thermocouple (Ellab A/S, Hilleroed, Denmark) and the temperature was read at display, with acoustic indicator at the end of the measurement.

9.4 Plasma Orexin A

The blood sampling was carried out in all participants. At 8:00 am, the blood sampling was carried out in all subjects, who were fasting from 8:00 pm. The blood derived from forearm vein and were utilized Vacutainer tubes (BD, Franklin Lakes, NJ, USA) containing EDTA and 0.45 TIU/mL of aprotinin. Blood samples were quickly centrifuged (3000 rpm at 4 °C for 12 min) and were refrigerated at –80 °C until analytical measurements. Plasma orexin A concentrations were determined with enzyme-linked immunoassay (ELISA), using kits of Phoenix Pharmaceuticals (USA). To extract plasma orexin A were utilized Sep-Pak C18 columns (Waters, Milford, MA, USA) before the measurements of concentration of orexin A. The activation of columns was obtained with 10 mL of methanol and 20 mL of H_2O. A sample of 1–2 mL was applied to the column and washed with 20 mL H_2O. The elution was obtained with 80% acetonitrile and the resulting volume was reduced to 400 mL under flow of nitrogen. The dry residue, obtained with evaporation by Speedvac (Savant Instruments, Holbrook, NY, USA), was dissolved in water and utilized for ELISA. Any cross reactivity of the antibody for orexin A (16–33), orexin B, agouti-related protein (83–132)-amide were not detected. The minimal detectable concentration was 0.37 ng/mL. The inter-assay error and intra-assay error were <14% and <5%, respectively.

9.5 Statistical Analysis

Data were analyzed using the GraphPad Prism 6 software for Windows (Microsoft, USA). The analysis of variance for repeated measures (ANOVA) was performed to determine differences among the dependent variables for the effect of training stage. When indicated by a significant F-value, a post hoc test using the Bonferroni multiple comparisons was used to identify significant differences between groups. Multivariate regression analysis was used to evaluate the role of confounding factors on results. All data were reported as means $\pm$ SD. Statistical significance was considered for $p \leq 0.05$.

9.6 Results

Figure 9.1 represents HR changes. The physical activity caused a change in HR with maximum level at time 30 min. Heart rate (beats/min) increased from a basal value of 73.2 ± 4 to 141.4 ± 5.5 at 30 min, and it returned almost to basal value at 60 min. The ANOVA demonstrate significant effects [F $(3, 18) = 93.42$, $p < 0.01$]. The post hoc test proved a difference between pre-exercise and post-exercise values at 30 min.

Figure 9.2 represents GSR changes. The physical activity caused a change in GSR with maximum level at time 30 min. GSR (μS) increased from a basal value of 1.21 ± 0.11 to 1.44 ± 0.11 at 30 min, and it decreased to 1.55 ± 0.09 at 60 min.

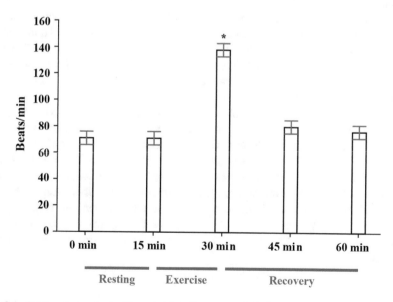

Fig. 9.1 Changes in heart rate. The asterisk indicates a statistical significant difference compared to other times

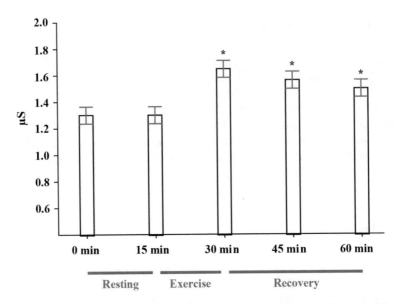

Fig. 9.2 Changes in galvanic skin response. The asterisk indicates a statistical significant difference compared to other times

The ANOVA proved significant effects [F $(3, 18) = 4.01$, $p < 0.05$]. The post hoc test proved a difference between pre-exercise and post-exercise values at 30, 45, and 60 min.

Figure 9.3 represents rectal temperature changes. The physical activity caused an increase in rectal temperature with maximum level at time 30 min. Rectal temperature (°C) raised from a basal value of 37.29 ± 0.08 to 37.75 ± 0.1, and it diminished to 37.59 ± 0.09 at 60 min. The ANOVA proved significant effects [F $(3, 18) = 4.98$, $p < 0.01$]. The post hoc test proved a difference between pre-exercise and post- exercise values at 30, 45, and 60 min.

Figure 9.4 represents orexin A changes. The physical activity caused an increase in orexin A from a basal value of 2617 ± 191 pg/mL to a maximum level (3566 ± 199 pg/mL) at time 30 min. Orexin A decreased to a value of 2961 ± 193 at 60 min. The ANOVA proved significant effects [F $(3, 18) = 5.07$, $p < 0.01$]. The post hoc test proved a difference between pre-exercise and post-exercise values at 30, 45, and 60 min. The same values on basal times (0 and 15 min) suggest that the monitored variables do not change over a short period. This justifies the lack of the control group. There was a normal distribution of all data. Multivariate regression analysis showed that there were no significant correlations between the variables examined.

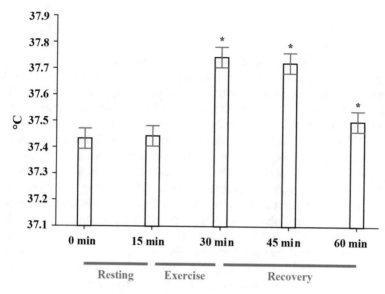

Fig. 9.3 Changes in rectal temperature. The asterisk indicates a statistical significant difference compared to other times

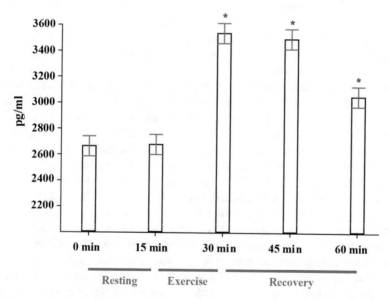

Fig. 9.4 Changes in plasmatic concentration of orexin A. The asterisk indicates a statistical significant difference compared to other times

9.7   Discussion

It is now widely known that sport induces psychological and physiological changes in the organism of those who practice it [22]. Sport is linked with the sympathetic activation [24], as demonstrated (even in this experiment) by a rise in the GSR and HR during physical activity. Exercise generates adaptive responses to sustain muscle engagement. Sport induces not only the sympathetic activation, but also hormonal changes. Furthermore, physical exercise is an effective way for enhancing cognitive performance and regulating mood. In healthy people, exercise reduces the likelihood of sleep and promotes alertness, whereas forced bed rest has the opposite effects [25, 26]. It also produced morphological and functional changes of brain regions that present an important role in successful everyday functioning, such as temporal and frontal cortices, and the hippocampus located in the inner (medial) region of the temporal lobe [27]. The factors most likely involved in exercise-induced hippocampal neurogenesis are the microcirculation and the production of neurotrophic factors such as the IGF-1, VEGF and BDNF [28]. Another probably factor that could contribute to the beneficial effects of sport is the orexin-A: exercise produces a rise in orexin-A level in cerebrospinal fluid of rats, dogs and cats [29]. Furthermore (as demonstrated in this experiment) physical exercise produces an increase of orexin-A level in plasma of humans. The increase of plasmatic orexin A in this experiment could be due to a rise in the secretion of orexin A by b-cells, in fact the pancreas is extensively innervated and secretion is regulated by neurotransmitter release. The increase in plasmatic orexin A could be included among hormonal adaptations due to exercise. Plasmatic orexin A is probably involved in the modulation of cortisol secretion. Indeed, relationship between orexin A and cortisol has been described. As orexin causes an enhancement of the expression of steroidogenic enzymes in human adrenocortical cells [30], an influence of orexin A on cortisol secretion, enhanced during physical activity, is conceivable. It is also worth pointing out that Sport is effective in the alleviation of depressive symptoms [31] and that a decrease of Orexin-A was found in animal models of depression and its intracerebroventricular administration reduced depression symptoms [32]. Is demonstrated that Sport helps to maintain a correct body weight and that transgenic mice with gradual and then loss of hypothalamic orexin-containing neurons shows feeding abnormalities and dysregulation in energy homeostasis determining obesity despite the reduction of food intake/calories [33]. Note that the orexin-A rapidly cross the blood- brain barrier, probably by simple diffusion, having a high degree of lipophilicity. Electrophysiological studies on transgenic mice have showed several neurotransmitters and neuromodulators influencing the activation or inhibition in orexin neurons activity and it is well known that cerebral orexin A is linked to the sympathetic discharge. An intracerebral administration of orexin A causes a widespread increase in the sympathetic discharge with augmented firing rate of sympathetic fibers innervating IBAT [34], kidneys [35], heart [36], etc. These evidences demonstrate that cerebral orexin system modulates the sympathetic nervous system with strong relationship of "cause-effect" between orexin A and sympathetic activation. In this experiment, the sympathetic discharge is positively associated with the level of plasmatic orexin A. Indeed, muscle exercise produces an increase in GRS and HR, both controlled by the sympathetic nervous system, which are linked by an increase in plasmatic orexin A. The results of our experiment are the first ones demonstrates a link between the plasmatic orexin A and the sympathetic activity but do not indicate a secure cause-effect relationship, but only an association. Despite this it is plausible that orexin A may influence the sympathetic activity and/or vice versa. The association between an increase in plasmatic orexin A and hyperthermia during physical activity, showed in the present experiment, may represent an involvement of peripheral orexin A in the control of body temperature during sport. Despite this evidence this experiment shows a modification of plasmatic orexin A during hyperthermia due to muscle exercise but its results are not enough to explain the role of orexin A in controlling the body temperature and sympathetic activity. Our study does not completely explain the role related to the elevation of orexin A level during exercise, but demonstrates a network between plasmatic orexin A and sport, both involved in the physiological perturbation induced by exercise. A lot more work is required to evaluate the role of orexin A in the sympathetic responses which coincide with exercise.

9.8   Conclusions

Sport is a backbone of wellness, because of the physiological perturbation induced by exercise and orexin is necessary for healthy life, because of the important and relevant homeostatic functions controlled and organized directly or not. Will be really important to explain whether the increase in sympathetic activity, during physical activity, precedes or follows the increase in levels of orexin, so as to fully understand the role of theoressinergic system in physical activity and in the proper functioning of the autonomic nervous system. Fully understand the network between orexin and sport could be a huge step towards understanding systems that are still little known and, that, could be extremely important to the scientific community given the many implications involved and the countless applications they might derive from.

9.9   Limitations

The main limitations of the present study are the serum level of the orexin A without the dosage on liquor; moreover, the measurement of sympathetic response using GRS, HR, Rectal Temperature rather than the dosage of catechalamines, but nevertheless, on the date there is no better non-invasive alternative to these techniques.

Conflict of Interest Statement   The authors declare that the research was conducted in the absence of any commercial or financial relationships that could be construed as a potential conflict of interest.

References

1. Nambu, T., Sakurai, T., Mizukami, K., Hosoya, Y., Yanagisawa, M., Goto, K.: Distribution of orexin neurons in the adult rat brain. Brain Res. **827**(1–2), 243–260 (1999)
2. Peyron, C., Tighe, D.K., Van Den Pol, A.N., De Lecea, L., Heller, H.C., Sutcliffe, J.G., Kilduff, T.S.: Neurons containing hypocretin (orexin) project to multiple neuronal systems. J. Neurosci. **18**(23), 9996–10015 (1998)
3. Chieffi, S., Messina, G., Villano, I., Messina, A., Esposito, M., Monda, V., Viggiano, A.: Exercise influence on hippocampal function: possible involvement of orexin-A. Front. Physiol. **8**, 85 (2017)
4. Messina, A., De Fusco, C., Monda, V., Esposito, M., Moscatelli, F., Valenzano, A., Cibelli, G.: Role of the orexin system on the hypothalamus-pituitary-thyroid axis. Front. Neural Circuits **10**, 66 (2016)
5. Viggiano, A., Chieffi, S., Tafuri, D., Messina, G., Monda, M., De Luca, B.: Laterality of a second player position affects lateral deviation of basketball shooting. J. Sports Sci. **32**(1), 46–52 (2014)
6. Messina, G., Monda, V., Moscatelli, F., Valenzano, A.A., Monda, G., Esposito, T., Cibelli, G.: Role of orexin system in obesity. Biol. Med. **7**(4), 1 (2015)
7. Messina, G., Dalia, C., Tafuri, D., Monda, V., Palmieri, F., Dato, A., Chieffi, S.: Orexin-A controls sympathetic activity and eating behavior. Front. Psychol. **5**, 997 (2014)
8. Hara, J., Beuckmann, C.T., Nambu, T., Willie, J.T., Chemelli, R.M., Sinton, C.M., Sakurai, T.: Genetic ablation of orexin neurons in mice results in narcolepsy, hypophagia, and obesity. Neuron **30**(2), 345–354 (2001)
9. Harris, G.C., Wimmer, M., Aston-Jones, G.: A role for lateral hypothalamic orexin neurons in reward seeking. Nature **437**(7058), 556 (2005)
10. Peyron, C., Faraco, J., Rogers, W., Ripley, B., Overeem, S., Charnay, Y., Li, R.: A mutation in a case of early onset narcolepsy and a generalized absence of hypocretin peptides in human narcoleptic brains. Nat. Med. **6**(9), 991 (2000)
11. Thannickal, T.C., Moore, R.Y., Nienhuis, R., Ramanathan, L., Gulyani, S., Aldrich, M., Siegel, J.M.: Reduced number of hypocretin neurons in human narcolepsy. Neuron **27**(3), 469–474 (2000)
12. Yamanaka, A., Beuckmann, C.T., Willie, J.T., Hara, J., Tsujino, N., Mieda, M., Yanagisawa, M.: Hypothalamic orexin neurons regulate arousal according to energy balance in mice. Neuron **38**(5), 701–713 (2003)
13. Messina, G., Zannella, C., Monda, V., Dato, A., Liccardo, D., De Blasio, S., Monda, M.: The beneficial effects of coffee in human nutrition. Biol. Med. **7**(4), 1 (2015)
14. Lu, X.Y., Bagnol, D., Burke, S., Akil, H., Watson, S.J.: Differential distribution and regulation of OX1 and OX2 orexin/hypocretin receptor messenger RNA in the brain upon fasting. Horm. Behav. **37**(4), 335–344 (2000)
15. Trivedi, P., Yu, H., MacNeil, D.J., Van der Ploeg, L.H.T., Guan, X.M.: Distribution of orexin receptor mRNA in the rat brain. FEBS Lett. **438**(1–2), 71–75 (1998)
16. Xu, T.R., Yang, Y., Ward, R., Gao, L., Liu, Y.: Orexin receptors: multi-functional therapeutic targets for sleeping disorders, eating disorders, drug addiction, cancers and other physiological disorders. Cell. Signal. **25**(12), 2413–2423 (2013)
17. Dall'Aglio, C., Pedini, V., Scocco, P., Boiti, C., Ceccarelli, P.: Immunohistochemical evidence of Orexin-A in the pancreatic beta cells of domestic animals. Res. Vet. Sci. **89**(2), 147–149 (2010)
18. Monda, M., Viggiano, A.N., Viggiano, A., Viggiano, E., Lanza, A., De Luca, V.: Hyperthermic reactions induced by orexin A: role of the ventromedial hypothalamus. Eur. J. Neurosci. **22**(5), 1169–1175 (2005)
19. Lubkin, M., Stricker-Krongrad, A.: Independent feeding and metabolic actions of orexins in mice. Biochem. Biophys. Res. Commun. **253**(2), 241–245 (1998)

20. Shirasaka, T., Nakazato, M., Matsukura, S., Takasaki, M., Kannan, H.: Sympathetic and cardiovascular actions of orexins in conscious rats. Am. J. Physiol.-Regul., Integr. Comp. Physiol. **277**(6), R1780–R1785 (1999)
21. Monda, M., Viggiano, A., Mondola, P., De Luca, V.: Inhibition of prostaglandin synthesis reduces hyperthermic reactions induced by hypocretin-1/orexin A. Brain Res. **909**(1–2), 68–74 (2001)
22. Zhang, W., Sakurai, T., Fukuda, Y., Kuwaki, T.: Orexin neuron-mediated skeletal muscle vasodilation and shift of baroreflex during defense response in mice. Am. J. Physiol.-Regul., Integr. Comp. Physiol. **290**(6), R1654–R1663 (2006)
23. Monda, M., Messina, G., Scognamiglio, I., Lombardi, A., Martin, G. A., Sperlongano, P., Stiuso, P.: Short-term diet and moderate exercise in young overweight men modulate cardiocyte and hepatocarcinoma survival by oxidative stress. Oxidative Med. Cell. longevity (2014)
24. Moscatelli, F., Valenzano, A., Petito, A., Triggiani, A.I., Ciliberti, M.A.P., Luongo, L., Monda, M.: Relationship between blood lactate and cortical excitability between taekwondo athletes and non-athletes after hand-grip exercise. Somatosens. Mot. Res. **33**(2), 137–144 (2016)
25. Bonnet, M.H., Arand, D.L.: Sleepiness as measured by modified multiple sleep latency testing varies as a function of preceding activity. Sleep **21**(5), 477–483 (1998)
26. Campbell, S.S.: Duration and placement of sleep in a "disentrained" environment. Psychophysiology **21**(1), 106–113 (1984)
27. Chieffi, S., Iachini, T., Iavarone, A., Messina, G., Viggiano, A., Monda, M.: Flanker interference effects in a line bisection task. Exp. Brain Res. **232**(4), 1327–1334 (2014)
28. Cotman, C.W., Berchtold, N.C., Christie, L.A.: Exercise builds brain health: key roles of growth factor cascades and inflammation. Trends Neurosci. **30**(9), 464–472 (2007)
29. Messina, G., Di Bernardo, G., Viggiano, A., De Luca, V., Monda, V., Messina, A., Monda, M.: Exercise increases the level of plasma orexin A in humans. J. Basic Clin. Physiol. Pharmacol. **27**(6), 611–616 (2016)
30. Wenzel, J., Grabinski, N., Knopp, C.A., Dendorfer, A., Ramanjaneya, M., Randeva, H.S., Jöhren, O.: Hypocretin/orexin increases the expression of steroidogenic enzymes in human adrenocortical NCI H295R cells. Am. J. Physiol.-Regul., Integr. Comp. Physiol. **297**(5), R1601–R1609 (2009)
31. Rethorst, C.D., Landers, D.M., Nagoshi, C.T., Ross, J.T.: Efficacy of exercise in reducing depressive symptoms across 5-HTTLPR genotypes. Med. Sci. Sports Exerc. **42**(11), 2141–2147 (2010)
32. Arendt, D.H., Ronan, P.J., Oliver, K.D., Callahan, L.B., Summers, T.R., Summers, C.H.: Depressive behavior and activation of the orexin/hypocretin system. Behav. Neurosci. **127**(1), 86 (2013)
33. Yokobori, E., Kojima, K., Azuma, M., Kang, K.S., Maejima, S., Uchiyama, M., Matsuda, K.: Stimulatory effect of intracerebroventricular administration of orexin A on food intake in the zebrafish. Danio rerio. Peptides **32**(7), 1357–1362 (2011)
34. Monda, M., Viggiano, A., Viggiano, A., Viggiano, E., Messina, G., Tafuri, D., De Luca, V.: Sympathetic and hyperthermic reactions by orexin A: role of cerebral catecholaminergic neurons. Regul. Pept. **139**(1–3), 39–44 (2007)
35. Monda, M., Sullo, A., De Luca, E., Pellicano, M.P.: Lysine acetylsalicylate modifies aphagia and thermogenic changes induced by lateral hypothalamic lesion. Am. J. Physiol.-Regul., Integr. Comp. Physiol. **271**(6), R1638–R1642 (1996)
36. Shahid, I.Z., Rahman, A.A., Pilowsky, P.M.: Orexin and central regulation of cardiorespiratory system. In: Vitamins & Hormones, vol. 89, pp. 159–184. Academic Press, New York (2012)
37. Sperandeo, R., Maldonato, N.M., Messina, A., et al.: Orexin system: network multi-tasking. Acta Medica Mediterranea **34**, 203 (2018)

Chapter 10
Experimental Analysis of in-Air Trajectories at Long Distances in Online Handwriting

Carlos Alonso-Martinez and **Marcos Faundez-Zanuy**

Abstract In this paper, we analyze in-air movements in online handwriting databases when the distance from the tip of the pen to the paper surface is higher than 1 cm. In this case, the computer can only know the time spent in air because the distance is too high to track the x and y coordinates of the movement. While this kind of movement is usually discarded, some investigation must be done in order to decide if computational algorithms can take advantage of this information in some scenarios. In this paper, we establish a criterion to differentiate useful in-air long distance strokes from user pauses.

Keywords Handwriting · Biometrics · In-air trajectories

10.1 Introduction

Signal analysis requires a set of samples. The larger the dataset, the more reliable the analysis is. However, experimental results are highly dependent on signal acquisition conditions. Multi-center and multisession signal acquisition, wide range of tasks and donor ages, etc. are desirable. However, data acquisition requires an enormous work and is time consuming. Fortunately, some public databases are available for research and this first step of data acquisition can be avoided. The main advantage of using public available databases, in addition to avoiding the load of data acquisition, is that experimental results can be replicated by other researchers if enough details are provided. In addition, comparison with previous published results is straighter forward. When using a new database, it is harder to compare because better accuracies can be motivated by the simplicity of the database rather than an improvement on the applied algorithms and techniques.

C. Alonso-Martinez · M. Faundez-Zanuy (✉)
ESUP Tecnocampus (UPF), Av. Ermest Lluch 32, 08302 Mataró, Spain
e-mail: faundez@tecnocampus.cat

C. Alonso-Martinez
e-mail: calonso@tecnocampus.cat

© Springer International Publishing AG, part of Springer Nature 2019
A. Esposito et al. (eds.), *Quantifying and Processing Biomedical and Behavioral Signals*, Smart Innovation, Systems and Technologies 103,
https://doi.org/10.1007/978-3-319-95095-2_10

109

In this paper, we use a large set of online handwritten available databases. The main advantage is that we can cover a wide range of situations, including medical pathologies such as dementia (Parkinson and Alzheimer), hypoxia, depression, etc., as well as tasks (signature, cursive and capital letters, drawings, etc.). The main drawback is that these databases have not been acquired thinking of in-air long distance movements and no specific instructions were provided to donors about this issue. Thus, we have found some users and samples where these movements are very long, and are not probably due to the task realization. They seem to be related to pauses, and probably conversations with the database acquisition supervisor in the middle of some tasks. Thus, a criterion must be established to differentiate those movements that are part of the handwriting task from those movements that are pauses in the middle of the task. Obviously, this last case is not very frequent, but due to the enormously high values, it has to be detected. Otherwise statistical averages can be biased due to this few samples, which must be considered outliers.

10.1.1   Database Description

We have analyzed a set of different databases that contain different tasks and user profiles. The databases share the existence of handwritten tasks.

Biosecur-ID

A multimodal biometric one which includes eight biometric traits: speech, iris, face (still images and videos), handwritten signature and handwritten text, fingerprints, hand and keystroking. This database acquired inside the BiosecurID project, was developed by a consortium of six Spanish Universities, you can find more details in Fierrez et al. [1]. With regard to handwriting and signatures, this database defines five different tasks: a Spanish text in lower-case, ten digits written separately, sixteen Spanish words in upper-case, four genuine signatures and one forgery of the three precedent subjects. These tasks were recorded for 400 subjects in four different sessions.

EMOTHAW

As described in [2], this database includes samples of 129 participants which are classified in base of their emotional states: anxiety, depression, and stress or healthy, this classification is assessed by the Depression–Anxiety–Stress Scales (DASS) questionnaire. Seven tasks are recorded through a digitizing tablet: pentagons and house drawing, words in capital letters copied in handprint, circles with left and right hand, clock drawing, and one sentence copied in cursive writing.

PaHaW

The Parkinson's Disease Handwriting Database (PaHaW) is built with multiple handwriting samples from 37 parkinsonian patients and 38 gender and age matched controls. Eight different tasks were recorded through a digitizing tablet: spiral draw-ing, letters, words, and a sentence. The details about this database can be found in [3].

10.1.2 Online Handwritten Signal

Online handwritten signal acquisition consists of sampling the movements performed by the user when moving the writing device. Digitizing tablet usually provides the following information:

(a) Absolute spatial coordinates (x, y) of the tip of the pen.
(b) Pressure exerted on the surface. Of course, this value is zero when the pen is not touching the surface.
(c) Angles of the pen: altitude and azimuth.
(d) Time stamp of the moment where the previous values have been acquired.

When pressure is different from zero the movement is considered to be on-surface and the whole set of information described before is acquired. When pressure is zero, the movement is considered to be in-air. If the distance from the tip of the pen to the paper surface is below one centimeter the whole set of information described before is acquired with the unique exception of pressure, which is always zero. However, when the distance is higher than 1 cm the track between the tablet and the pen is lost and no samples are acquired. This means that the previous described information (a)–(d) is not acquired. However, as soon as the pen touches the surface, samples are acquired again. Working out the difference between consecutive time stamps we can detect the time spent at this long-distance movements.

10.2 Outlier Detection

During capturing process, some not desired situations can produce undesired untracked intervals. In order to study in-air long distance movements, it is neces-sary to remove this outlier values. In Fig. 10.1 an example of handwritten task is shown with its time stamp difference bar diagram. We can observe that in-air long distance intervals are lower than 0.03 s which it is consistent with the normal writing process.

In Fig. 10.2 we can see the same task from another subject, but in this case, an outlier value is shown in the time stamp difference graph, because an interval of approximately seven seconds not seem to be consistent with a normal writing process.

The question is to decide which intervals can be considered atypical values and which are not. In a first approximation, we will study the percentage of the total time of the file corresponding to these in-air long distance intervals. This allows, to

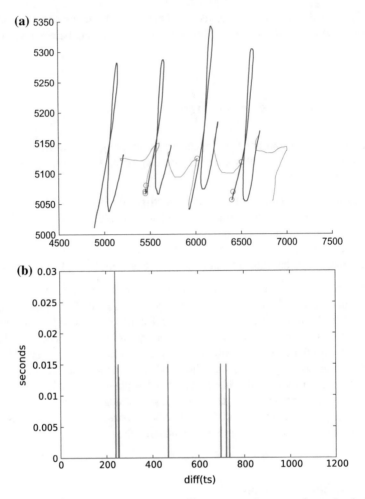

Fig. 10.1 **a**. Handwritten task example: continued line corresponds to on-surface traits, dashed line to in-air near distance traits and circle to in-air long distance gaps. **b**. Time stamp difference from handwritten task example

detect cases where pauses in writing appear not related to the natural movements of the writing, but to anomalous situations, such as that where the subject pauses or interrupts the task to talk. It is used as reference values for the study, situations where the time in long distance, exceeds 70, 50, 40 and 30% respectively, of the total time of the file.

In Table 10.1 the results of BiosecurId database are presented, indicating the number and percentage of files which duration exceed the threshold.

Firstly, it is observed how few files have atypical values for in-air long distance intervals, can be emphasized the task of writing in capital letters, where in 0.08% of the files, the total in-air long distance time exceeds the 30% of full file length.

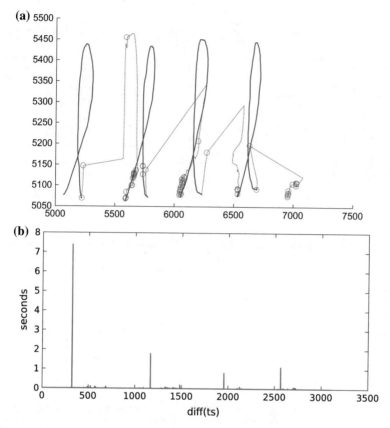

Fig. 10.2   a. Handwritten task example with outliers: continued line corresponds to on-surface traits, dashed line to in-air near distance traits and circle to in-air long distance gaps. **b.** Time stamp difference from handwritten task example with outliers

Table 10.1   BiosecurId: files by task exceeding the in-air long distance threshold

Task	50%	40%	30%
Paragraph	1 (0.063%)	2 (0.125%)	5 (0.313%)
Numbers	0	0	4 (0.250%)
Words capital	0	2 (0.125%)	13 (0.813%)
Genuine signatures	0	1 (0.016%)	2 (0.0.31%)
Forgery signatures	0	4 (0.083%)	11 (0.229%)

Results for PaHaW database are shown for control subjects (see Table 10.2) and for Parkinson patients (see Table 10.3).

As for a base with many less subjects, the effect of even a single file is much greater, therefore, the detection of these atypical values is even more important. For example, the file corresponding to a phrase that exceeds 70% threshold presents an

Table 10.2 PaHaW: files by task exceeding the in-air long distance threshold for healthy subjects

Task	50%	40%	30%
Spiral	0	0	0
Letter l five times	0	0	2 (5.263%)
Bigram le	0	0	1 (2.632%)
Les word	0	0	0
Lektorka word	1 (2.632%)	1 (2.632%)	2 (5.263%)
Porovnat word	0	0	0
Nepopadnout word	0	0	0
Phrasse	0	0	0

Table 10.3 PaHaW: files by task exceeding the in-air long distance threshold for Parkinson patients

Task	70%	50%	40%	30%
Spiral	0	0	0	0
Letter l five times	0	0	0	0
Bigram le	0	0	0	0
Les word	0	0	1 (2.703%)	2 (5.405%)
Lektorka word	0	0	0	0
Porovnat word	0	0	0	0
Nepopadnout word	0	0	0	0
Phrasse	1 (2.703%)	1 (2.703%)	2 (5.405%)	2 (5.405%)

abnormal gap in the time stamp values at the end of the file (see Fig. 10.3) which it is clearly an error in the time stamp recording procedure and has no relation with the movements of the grapho-scriptural task.

EMOTHAW database results are shown in Table 10.4. It is remarkable as the high percentage of cases for the clock drawing task where in-air long distance time exceeds the 30% of total length of file. For this reason, it does not seem reasonable to dismiss this cases as atypical values.

Once the existence of anomalous values has been verified in the databases, the next step is to study how to eliminate these outliers. There are several formulas for deciding which values are considered outliers [4]. One of the most widely used, is based interquartile range (IQR), determining that values higher than $Q3 + 1.5$ IQR, where Q3 corresponds to the third quartile and IQR $= Q3 - Q1$.

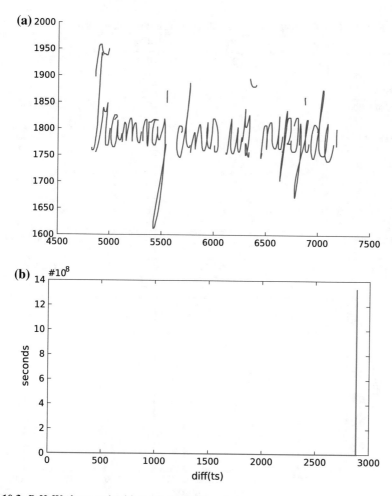

Fig. 10.3 PaHaW phrase task with atypical large timestamp gap

The first choice would be calculating intra-file threshold, that is, determine for every file the in-air long distance intervals threshold and eliminate outliers. The drawback of this method is that eliminate values that although elevated, are characteristics of the drawing or writing process and therefore can provide information. In Fig. 10.4 the effect of an intra-file threshold can be observed for a clock drawing task (EMOTHAW database) and how some interesting in-air long distance interval are eliminated.

For this reason, we propose calculate inter-file threshold for every task of every database and obtain a threshold which can be used for eliminate outliers. The procedure is similar to the previous one, but in this case, we take the maximum values for every file and then, we evaluate with these values the threshold, Q3 + 1.5 IQR. In Tables 10.5, 10.6 and 10.7 upper thresholds are shown for the different databases.

Table 10.4 EMOTHAW: files by task exceeding the in-air long distance threshold

Task	70%	50%	40%	30%
Copy of a two-pentagon drawing	0	0	0	5 (3.876%)
Copy of a house drawing	0	0	0	2 (1.550%)
Writing of four Italian words in capital letters	0	0	3 (2.326%)	4 (3.101%)
Loops with left hand	0	0	0	4 (3.101%)
Loops with right hand	0	0	0	0
Clock drawing test	0	0	0	19 (14.729%)
Sentence	1 (0.775%)	1 (0.775%	1 (0.775%	2 (1.550%)

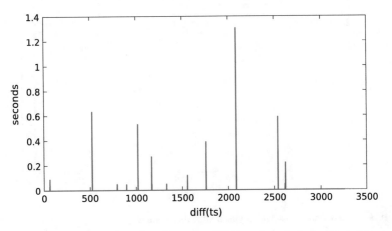

Fig. 10.4 Pre and post intra-file outliers time stamp difference for a clock drawing task

Table 10.5 BiosecurId inter-file thresholds

Task	Threshold (ms)
Paragraph	4008.5
Numbers	60
Words capital	4669.25
Genuine signatures	52.5
Forgery Signatures	65

Table 10.6 PaHaW inter-file thresholds

Task	Threshold (ms)
Espiral	56.875
Letter l	765.75
Bigram le	1152.25
Les	1288.125
Lektorka	2023.875
Porovnat	953.625
Nepopadnout	103.25
Phrasse	4042.625

Table 10.7 EMOTHAW inter-file thresholds

Task	Threshold (ms)
Copy of a two-pentagon drawing	1385.25
Copy of a house drawing	2188.75
Writing of four Italian words in capital letters	3656.25
Loops with left hand	0
Loops with right hand	0
Clock drawing test	3680.5
Sentence	1640.25

10.3 Conclusions

The study of values implies the need to eliminate outliers since in the processes of registration, errors and incidences occur that affects the timestamp. Although in small numbers, as has been observed, can distort the results and therefore, it is necessary to eliminate them as a previous step.

The use of inter-file thresholds allows to eliminate anomalous values and at the same time maintain the in-air long distance intervals, characteristics of the writing or drawing process.

Acknowledgements This work has been supported by FEDER and MEC, TEC2016-77791-C4-2-R.

References

1. Fierrez, J., et al.: BiosecurID: a multimodal biometric database. Patt. Anal. Appl. **13**(2), 235–246 (2010)
2. Likforman-Sulem, L., Esposito, A., Faundez-Zanuy, M., Clémençon, S., Cordasco, G.E.: A novel database for emotional state recognition from handwriting and drawing. IEEE Trans. Hum.-Mach. Syst. **99**, 1–12 (2017). https://doi.org/10.1109/thms.2016.2635441

3. Drotár, P., Mekyska, J., Rektorová, I., Masarová, L., Smékal, Z., Faundez-Zanuy, M.: Evaluation of handwriting kinematics and pressure for differential diagnosis of Parkinson's disease. Artif. Intell. Med. **67**, 39–46 (2016)
4. Montgomery, D., et al.: Probabilidad y estadística aplicadas a la ingeniería. McGraw Hill, New York (1996)

Chapter 11
Multi-sensor Database for Cervical Area: Inertial, EEG and Thermography Data

Xavi Font, Carles Paul and Joan Moreno

Abstract Inertial sensors for analysing some biomechanics conditions have long been studied in the medical and sports fields. In order to improve any qualitative assessment related to detecting the degree of injury and the range of motion in the cervical area, the system needs to be robust enough. It is important to detect purposefully altered data, reduce subjective variables effect (such as pain or discomfort) and accurately determine the impairment or dysfunction levels. The first aim of this work was to produce a multi-sensor database for the cervical area by gathering data from an inertial system, from an EEG head-set and from a thermographic camera. This complete set of information can provide further insight when researchers try to develop an objective diagnostic algorithm or improve intelligent diagnostic systems.

Keywords Biomechanical experimentation · Injury detection · Improving diagnostic algorithms · Thermal images · Inertial data · EEG data

11.1 Introduction

The task of building a system which can assist with decision-making for the clinical assessment of pathologies affecting the musculoskeletal system is a challenging problem [1, 2]. The options available and the approach that one can take are wide enough to understand that there is not an optimal methodology to deal with it.

It is understandable that insurance companies and public health systems need to assess the level of functionality lost in order to provide a correct rehabilitation strategy and to accurately cover the expenses related to the injury.

X. Font (✉) · C. Paul · J. Moreno
Tecnocampus-UPF, Mataro, Barcelona, Spain
e-mail: font@tecnocampus.cat

C. Paul
e-mail: paul@tecnocampus.cat

J. Moreno
e-mail: joan.moreno.mamano@eupmt.tecnocampus.cat

© Springer International Publishing AG, part of Springer Nature 2019   119
A. Esposito et al. (eds.), *Quantifying and Processing Biomedical and Behavioral Signals*, Smart Innovation, Systems and Technologies 103,
https://doi.org/10.1007/978-3-319-95095-2_11

The main aim of this work is to provide a new data base that goes a step further with the addition of three different sources of information. That is, data coming from an inertial system, data acquired from a thermographic camera and data provided by an EEG head-set. Some research has been made [3] for specific sensors.

11.2   Experiment Design

The experiment was designed according to certain requirements, which means obtaining the range of motion, and velocity in a three-step exercise. These three steps of the exercise can be described as:

- Extension/flexion (or in other words the *Yes* - movement)
- Inclination, left and right (*I do not know* - movement)
- Rotation, left and right (*No* - movement)

thus all users will go through these three steps as shown in Fig. 11.1.

Before starting the exercise, a set of thermographic shots $(20 + 20 + 20)$ will be programmed to cover the left, the right side of the head and the back as well. This process is repeated after the exercise. Data will flow through the H2620 NEC camera (5 s/shot) The inertial data and the EEG data will be acquired while users repeat the described exercise 10 times.

All users go through the same process. First they signed an agreement to authorise the transfer of personal details in accordance with the Spanish Organic Law for scientific purposes and then they completed a survey to give additional information such as age, gender, injury level, degree of pain, headaches and so on. In total 27 users ranging from 20 years old to 60 years old provide of their time and data generously.

The complete process lasted almost 40 minutes per user. During this time all the users completed the same process. Some pictures of the process are shown in Fig. 11.2. It is worth mentioning that at least two acquisitions where obtained for each user.

11.2.1   Subjects

Twenty seven subjects were enrolled with ages ranging from 20 to 60 years old. There were 14 subjects in good health(no injury present) and 13 with some level of injury. The complete set of users are shown in Table 11.1.

Fig. 11.1 Movements (from left to right): Yes ; I don't Know ; No

Fig. 11.2 From left to right: thermographic acquisition process, lab acquisition area and Emotiv + EEG head-set

Table 11.1 Subjects enrolled in the multi-sensor database for cervical area

	Subjects with injury	Healthy subjects	Total
20–40 Years	4 Women + 3 Men	4 Women + 3 Men	*8 Women + 6 Men*
41–60 Years	3 Women + 3 Men	3 Women + 4 Men	*6 Women + 7 Men*
Total	*7 Women + 6 Men*	*7 Women + 7 Men*	**14 Women + 13 Men**

No healthy user complained about the exercises, however some injured subjects expressed their discomfort at the end of the process. All this data was collected via a survey to help exploit the data acquired from the three different sensors: inertial, thermography and EEG.

11.3 Inertial Data

Inertial data is one of the main sources of information for clinicians, sports professionals and researchers alike [4]. Inertial sensors are sensors based on inertia using magnetometers (accelerometer and gyroscope) that gives reasonable accuracy. These data is in many cases the only available information to asses the degree of impairment, or to quantify normal movements versus pathological movements [5, 6]. The innovation in sensing system is not enough to use the data for diagnosis. A signal processing work is needed or a software had to be used to obtain measures related to the area under study.

11.3.1 Acquisition Process

Before starting the exercise the user was sitting in a chair. A full description of the exercise was given while the sensors were placed in the correct positions and proved to be working properly. The first sensor is placed at T1 level (adhesive attachment) and the second sensor is located at the occipital bone (elastic tape see Fig. 11.3). The user repeated a "yes" movement 10 times (extension/flexion) then they continued with 10 additional repetitions with the "I don't know" movement (inclination Left/Right)

Fig. 11.3 From left to right: inertial sensor location, "Yes" + "I don't know" and "No" movements

and the final step was to repeat 10 "No" movements (rotation Left/Right). All these three exercise were done in succession if all the data coming from the inertial sensors was correct. It is worth mentioning that randomization would certainly increase the information about the effect of the three types of the exercises on the response variable. However, due to subject/patient availability and time restrictions, experimental design was kept to a minimum.

The complete set of data obtained through these sensors was:

- Time.
- Extension/Flexion, Inclination Left/Right and Rotation Left/Right.
- Extension/Flexion Velocity, Inclination Left/Right Velocity and Rotation Left/Right Velocity.
- Extension/Flexion Acceleration, Inclination Left/Right Acceleration and Rotation Left/Right Acceleration.

The units of the sensors were degrees, degrees/s and degrees/s^2. Figure 11.4 shows the output of the second, third and fourth column from some random users in the database.

It is not difficult to obtain a biomechanical report through the position analysis (variables related to extension/flexion, inclination L/R and rotation L/R) One step further might be possible correlating position and velocity (note three variables for positioning and three for velocity). Later on this data can be merged with the survey data to identify possible problems [7]

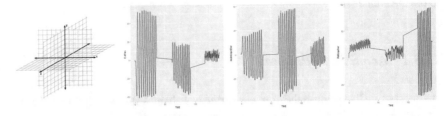

Fig. 11.4 From left to right: XYZ sensor axis, "Yes" + "I don't know" and "No" outputs

11.4 Thermographic Data

The use of thermography in the industry sector has a long history and currently plays an important role in validating normal operations. The key element to understanding why thermographic cameras are so important is with pinpointing thermal anomalies, and identify heat patterns that are related to possible defects, faults, inefficiencies and so on. However in this database the role of the thermographic images is related to body temperature and its link to disease [8]. The data tries to give an additional axis to study movement performance and possible injuries.

The equipment used was an Infrared Thermography H2640 from Nippon avionics CO.LTD. which has the following characteristics:

- High resolution 640 × 480-pixel detector.
- Resolution: 0.06 °C or better (at 30°, 30 Hz), 0.03 °C.
- Accuracy: ±2% (of reading) or ±2 °C.
- IFOV (Instantaneous Field of View or Spatial Resolution): 0.6 mrad.

Once the data was obtained, it was necessary to use the software *ReportGen* from NEC Avio Infrared Technologies to export the complete set of pictures to a more suitable data type. First the data was exported to an excel data type, and later an additional transformation was necessary to present the data in a format-friendly way for image processing. At the end there were 20 pictures for each user and repetition from the right, an additional 20 from the back and 20 more from the left side (see Fig. 11.5).

11.4.1 Acquisition Process

As previously mentioned, the first data acquired from the database was the one obtained before the user got started (10 "YES" repetition movement). A set of 20 shots were taken place from the three studied areas: right, back and left. The process was repeated after the subject finished the whole set of exercises.

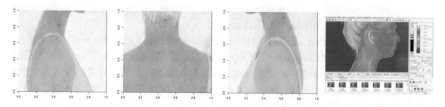

Fig. 11.5 From left to right: thermography with a hot spot from the right, back and left cervical area before exercise. ReportGen software

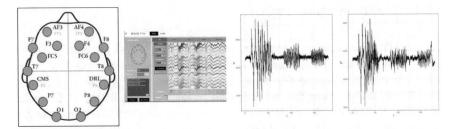

Fig. 11.6 From left to right: sensor location, emotiv acquisition software and two outputs from two channels (same user)

Again it is not difficult to make the most of image processing tools to obtain additional information from the thermographic images, lets say: hot spot, distribution of temperatures around the hot spot location, and so on.

11.5 EEG Data

The last sensor used to acquire data was an EEG headset (from EMOTIV+) The use of this additional input to obtain the electrical brain activity is a plus, and helps us to see a complete picture of each of the users enrolled on the database.

Without being given an insight into EEG [9], the data obtained throughout the exercise is related to the 14 channels of the device. Some of the data provided correlates with the one provided by the inertial sensors (see Fig. 11.6). This data was acquired at the same time as the one proceeding from the inertial sensors.

11.6 Conclusions

Based on the fact that it is the first multi-sensor database for the cervical area, with an easy biomechanical capture, the results that some can expect from it are very positive. The use of advanced machine learning methods over the database can bring insights to the problems related with cervical injuries.

An additional concern has to be stated. Unfortunately, data base accessibility (in the sense of an open data project) is not allowed at the moment. There are certain privacy claims that avoid sharing certain data without a proper de-identification process [10]. That being said, there is a partial data base that can be shared through a basic request by any of the authors.

Acknowledgements This work has been supported by FEDER and MEC, TEC2016-77791-C4-2-R.

References

1. Cuesta-Vargas, A.I., Galan-Mercant, A., Williams, J.M.: The use of inertial sensors system for human motion analysis. Phys. Ther. Rev. **15**, 462–473 (Dec 2010)
2. Voinea, G.-D., Butnariu, S., Mogan, G.: Measurement and Geometric Modelling of Human Spine Posture for Medical Rehabilitation Purposes Using a Wearable Monitoring System Based on Inertial Sensors. Sensors **17**, 0003 (Dec 2016)
3. Ngo, T.T., Makihara, Y., Nagahara, H., Mukaigawa, Y., Yagi, Y.: The largest inertial sensor-based gait database and performance evaluation of gait-based personal authentication. Pattern Recogn. **47**(1), 228–237 (2014)
4. Saber-Sheikh, K., Bryant, E.C., Glazzard, C., Hamel, A., Lee, R.Y.: Feasibility of using inertial sensors to assess human movement. Manual Ther. **15**, 122–125 (Feb 2010)
5. Theobald, P.S., Jones, M.D., Williams, J.M.: Do inertial sensors represent a viable method to reliably measure cervical spine range of motion? Manual Ther. **17**(1), 92–96 (2012)
6. A. N. A. Kuzmanic, T. Vlak, and I. V. O. Stancic, "Inertial Sensor Measurement of Head-Cervical Range of Motion in Transverse Plane," Data Process. **November 2016**, pp. 47–51, 2007
7. Huelke, D.F., O'Day, J., Mendelsohn, R.A.: Cervical injuries suffered in automobile crashes. J. Neurosurg. **54**(3), 316–322 (1981)
8. Feldman, F., Nickoloff, E.L.: Normal thermographic standards for the cervical spine and upper extremities. Skeletal Radiol. **12**, 235–249 (Nov 1984)
9. Park, J.L., Fairweather, M.M., Donaldson, D.I.: Making the case for mobile cognition: EEG and sports performance. Neurosci. Biobehav. Rev. **52**, 117–130 (May 2015)
10. Ribaric, S., Ariyaeeinia, A., Pavesic, N.: De-identification for privacy protection in multimedia content: a survey. Sig. Process. Image Commun. **47**, 131–151 (2016)

Chapter 12
Consciousness and the Archipelago of Functional Integration: On the Relation Between the Midbrain and the Ascending Reticular Activating System

Nelson Mauro Maldonato, Anna Esposito and Silvia Dell'Orco

Abstract Historically, the relation between consciousness and the brainstem has been demonstrated, on the one hand, by injuries to the upper brainstem that lead to minimum states of consciousness, comas and persistent vegetative states and, on the other hand, by electrophyisiological recordings that link the ascending reticular activating system (ARAS) with vigilance and attention, functions which are necessary for interpersonal relationships. With the advances made in the clarification of the connections between the brainstem and other regions of the brain there has been no corresponding conceptual revision of the functional context within which the ARAS performs its role of activation and way in which it activates the cerebral cortex, unlike other structures of the brain. In this paper we shall discuss the way in which the brainstem—(a) fundamental terminal of multiple ascending neural pathways—influences and modulates cortical activities; (b) the context of which the ARAS is a fundamental part—the centroencephalic archipelago of functional integration—for the transmission of contents to specific regions that generate the sensation of subjectivity.

N. M. Maldonato (✉)
Dipartimento di Neuroscienze e Scienze Riproduttive ed Odontostomatologiche,
Università di Napoli Federico II, Naples, Italy
e-mail: nelsonmauro.maldonato@unina.it

A. Esposito
Dipartimento di Psicologia, Università della Campania Luigi Vanvitelli, Caserta, Italy
e-mail: iiass.annaesp@tin.it

A. Esposito
IIASS, Vietri sul Mare, Italy

S. Dell'Orco
Dipartimento di Scienze Umane, Università della Basilicata, Potenza, Italy
e-mail: silvia.dellorco@unibas.it

© Springer International Publishing AG, part of Springer Nature 2019
A. Esposito et al. (eds.), *Quantifying and Processing Biomedical and Behavioral Signals*, Smart Innovation, Systems and Technologies 103,
https://doi.org/10.1007/978-3-319-95095-2_12

Keywords Consciousness · Ascending reticular activating system · Cortex
Brainstem · Highest level · Serotonin · Dopamine · Noradrenaline

12.1 Introduction

For a long time it was believed that consciousness was an expression of the cerebral
cortex, the most recent and elevated structure of the human brain. On the other hand,
it was believed that the emotional and instinctive functions were an expression of
the subcortical areas, the limbic system and the brain stem, the oldest structures with
automatic, instinctive and unconscious characteristics [1]. This point of view long
dominated (and to some extent still dominates) the scientific debate on consciousness,
despite (a) the inadequacy of a term that includes in its semantic field things that are
very different from each other [2]; (b) clinical evidence that shows how cortical
lesions, though significant, do not lead to a loss of consciousness [3]; (c) the fact that
the ablation of even large areas of the cerebral cortex does not cause disturbances
in consciousness; (d) the fact that epileptic fits have substantially little influence on
consciousness, apart from when they affect the centroencephalic system [4]; (e) the
fact that the isolation of a cortical motor area from those areas adjoining it does
not suppress the execution of a voluntary act [5]; (f) the fact that the presence of
particular physiopathological activities (an alpha rhythm or a convulsion) does not
suspend consciousness [6]. Today various studies confirm that single bilateral lesions
of some nuclei of the brainstem are not sufficient to cause comas [7].

12.2 Neuronal Maps

Around the middle of the last century, Moruzzi and Magoun [8] pointed out the
existence of a component of the brainstem capable of modulating, according to the
frequency and the intensity of the electrical stimulation, the global activity of the
cerebral cortex [9]. In particular, they observed that the stimulation of areas of the
brainstem provokes the reawakening of a sleeping animal and a state of alarm in
one that is awake [10]. It should be noted that, at an anatomic level, the brainstem
is organised in nuclei, that is, aggregates of neuronal bodies that are part of the grey
matter. Such groups are associated with the descending and ascending projective systems that traverse the brainstem [11]. Each nucleus has a prevalent neurotransmitter
specificity, its own localisation within the brainstem, its own connections with other
neural structures and, finally, a prevailing function. [12] These areas, situated in the
centro-dorsal regions of the stem—which host the reticular formation and extend
until the pons and the midbrain [13]—if affected by lesions of different kinds, can
cause comas or a persistent vegetative state in the person [14]. Therefore, both for the
localisation, and for the variety of connections with different functional meanings,
the reticular formation recognises its own specificity. In the last half-century, indeed,

various pieces of evidence have contributed to the modification of how it has been traditionally represented [15]. It has been seen, in fact that (a) it is not a homogeneous neuronal structure, but an anatomo-functional set of different nuclei; (b) each of its components may separately modulate the activity of the cerebral cortex; (c) the heterogeneity of its nuclei allows for the modulation of the activity of the cerebral cortex through pathways different from those of the intralaminar nuclei of the thalamus [16]. In this sense, some nuclei can influence the entire cortex through projections on telencephalic nuclei, while others can directly reach the cerebral hemispheres, thus having a modulatory effect [17]. Furthermore, some nuclei can modulate the electrophysiological activity of the cerebral cortex by modifying the activity of the reticular nucleus of the thalamus [18]. It is not without importance, therefore, that various ascending pathways from the reticular formation can utilise different neurotransmitters, thus modulating the cortical activity through feedback and feed forward mechanisms [19].

12.3 Transitive States and Substantive States

Even though the exact role of each ascending reticular nucleus is still quite unclear, it appears to be accepted that different nuclei and different pathways are involved in the modulation of the electrophysiological activity of the cerebral cortex [20]. In the light of these considerations, the activation of a system of nuclei with thalamocortical projections responsible for the electrophysiological cortical modulation, but which also projects towards the lower brainstem, could depend not so much on a single brainstem nucleus or on a set of nuclei, but on a network of different sets of nuclei [21]. Confirming this is the fact that the modulation of the global cerebral activity determined by the reticular formation is not limited to the midbrain, but affects many nuclei of the pons [22].

But in what way are such influences created in an ascending direction? Predominantly through two pathways: one extrathalamic, responsible for the awakening reactions; the other thalamic, which modulates the level of vigilance by orienting it in one direction rather than in another [23]. The latter pathway would be correlated, more than the first, with the attention which is a selective phenomenon and relates to the highest levels of nerve integration. What guarantees the efficiency of attention—and in general the functioning of the entire psychic apparatus—is vigilance [24]. Indeed, if the subject was not alert, not only the attention but also other faculties such as the memory, decision-making or motivation would remain in the potential state. As various electroencephalographic studies have shown, vigilance is sustained by oscillatory phenomena of parallel synchronisation and inhibition. These are transitive states and substantive states, the former characterised by unstable and high energy neuronal activity and the latter by stable and low energy neuronal activity [25]. In this sense, just as the reflex represents the lowest level of motor integration, vigilance represents the fundamental level of upper nerve integration that, indeed, has the ARAS as its basic structure [26].

12.4 The Highest Level

Around the middle of the last century, Penfield attributed to the centroencephalic system the function of highest level in the organisation of the superior functions, going so far as to claim that the control centre of consciousness resides in the high brainstem [27, 28]. As far as we know today, in the organisation of the central nervous system the central importance of the ARAS does not correspond to hierarchical superiority with respect to the hemispheres, which for their extremely high level of specialisation and greater selectivity perform their own at a different functional level [29]. On the other hand, for precise anatomical and physiological reasons, a specific projection system (olfactory, gustative, visual, aural or somesthetic) has functions that are quite different from another with a diffuse projection system. Indeed, the first is constituted by peripheral receptors, afferent pathways and relay nuclei of these same pathways, which reach a small portion of the cortex; the second by cellular aggregates distributed in the brainstem structures, the stimulation of which broadly influences the electrical activity of the brain [30]. Therefore, for its projections ascending towards the cerebral hemispheres and descending towards the lower brainstem [31], the ARAS performs functions that are considerably more expansive and complex than cortical desynchronisation alone, which is essential for alertness, attention and in other ways, creativity [32].

The hypothesis that we advance here is that the distributed and dynamic presence of "critical nuclei" of the brainstem [33] with their aspecific and high frequency oscillating thalamocortical projections are fundamental for some functions of consciousness. They are fundamental for their intimate connection with the hypothalamus and the main structures involved in the regulation of the primary emotional and vegetative life of the organism, to the creation of the body scheme, to movement and other fundamental systems. On the other hand, that the nuclei of the reticular formation contribute to modulate cortical activities is demonstrated by the intense relations with the principal neurotransmitter systems: serotonergic, noradrenergic and dopaminergic. The serotonergic system is involved in processes such as aggression, anxiety, hunger, memory, mood, sleep and thermoregulation [34]. The noradrenergic system is involved in the attentional processes and in the response to 'fight or flight' through the activation of the sympathetic nervous system that increases the heartbeat, releases energy in the form of glucose from glycogen and increases muscular tone [35]. The dopaminergic system is involved in behaviour, cognition, voluntary movement, motivation, mood, attention, and in the memory of work and learning [36]. These neurotransmitters intervene in the coordination of behavioural strategies in order to deal with environmental needs and to achieve homeostatic balance [37].

12.5 The Space of Centrencephalic Functional Integration

But in what way does the brain create neural models that map, instant by instant, the state of the organism in its multiple qualitative sensorial dimensions: smell, vision, hearing and so on? How does it generate representations and images that transfer the physical characteristics of an object to a wide network of relations with other objects, enhancing it with emotional-affective elements [38]? Finally, what is the nature of this spatio-temporal sequencing of the images within the sensory systems? Just as at certain levels the brain forms images of objects, at other levels it gives rise to events that are non-verbal and unaware of object-organism events [39]. Such dynamics vary depending on the functional organisation: it is relatively low when the connections are distributed statistically and is, however, at a maximum when it is connected to definite neuronal groups [40]. The higher the reciprocal information between each subset and the rest of the system, the greater the complexity. At the origin of the conscious experience there could be a distributed neuronal aggregate that performs operations of a duration of a few hundredths of a millisecond [41]. The supplementary activities necessary to superior consciousness could take place on the boundary between the thalamocortical system and the other areas of the brain [42]. In this space of functional integration, at the same time unified and separate, those remote correlations between different regions of the brain could take place, variable from moment to moment, at the origin of that form of primary cognition—non-aware, non-verbal and an expression of ancient phylogenetic and ontogenetic memories—that identifies us with our body scheme [43] and demarcates the borders between the self and the objects of external spatio-temporal contexts. In this model, the brain centre constitutes the neuronal infrastructure of the field verticality of consciousness [44] and the cerebral cortex becomes the place of the optional exercises of the field of consciousness.

12.6 That Unexplored and Silent Zone

In recent decades an important contribution has been made from the study of cerebral activation (fMR, PET, MEG, event-related potentials) that, in general anaesthesia, allow for the exploration of cerebral relations in the transition from general anaesthetic to reawakening, from the vegetative to the minimally conscious state and so on [45]. For example, the recovery of the thalamocortical activity in a patient who is initially in a vegetative state and then minimally conscious confirms, on the one hand, the importance of such connections in the processes of consciousness and, on the other hand, that this is not limited to a specific region of the brain [46]. The studies of the transition from oblivion to awareness carried out in within the sphere of anaesthesiology with the help of cerebral imaging currently offer important contributions to the knowledge of the conscious processes [47]. Recent research has shown how the reawakening from anaesthesia—with the full recovery of clear and oriented awareness—is often preceded by a strong turbulence in its contents [48]. Apart from the questions that are still unanswered, the further demonstration of the very strong co-implication of cortical and sub-cortical areas in the birth of consciousness is an important step forward. Experiments conducted by a team of researchers led by Harry Scheinin [49] on a sample of 20 young volunteers coming round from general anaesthetic have shown how the recovery of consciousness is correlated to an activation of centroencephalic structures, more than to the cerebral cortex. In other words, the state of awareness is preceded by the intensification of the metabolic activities of the centroencephalic nuclei, by the thalamus, the limbic system and by the lower frontoparietal areas [50]. The emergence of progressive levels of consciousness, evaluated according to the response to vocal commands, up to the highest level of consciousness of one's self and the world, initiates from these neural territories [51].

12.7 Conclusions

Over and above the debate on the role of thalamocortical oscillations, it seems to be accepted that at the origin of consciousness there is the simultaneous and distributed action of neural populations of the brainstem and of the cortical-subcortical axis. The brainstem represents the point of departure of ascending neural pathways which originate in distinct series of nuclei. After having reached the cortex, directly or by means of the thalamus and the cerebrum, these influence the cerebral cortex, both modulating aspects of its overall activity and transmitting the contents to specific regions through which a subjective sense can be generated. The continuous mapping and remapping [52] of the state of the organism creates functional contexts for the encephalic trunk nuclei that can modulate the functioning of the frontal cortex. Moreover, the modifications deriving from the interaction with ever-changing objects change the state of the organism in relation to causal objects as a base for conscious experience. It is plausible that this dynamic balance, in which each event (a visual image or something else entirely) reflects the activation of a neural network, distributed and in parallel, gives rise to stable contents of consciousness, as a precondition for the emergence of awareness. It is from this point of departure that in the future [53] research will be required to explore the difficult problem of consciousness.

References

1. Maldonato, N.M.: From neuron to consciousness: for an experience-based neuroscience. World Futures **65**(2), 80–93 (2009)
2. Maldonato N. M. Embodied Consciousness. World Futures, Special Issue, 2(65) (2009)
3. Koch, C.: The Quest for Consciousness: A Neurobiological Approach. Roberts and Company Publishers, Englewood, CO (2004)

4. Penfield, W.: The Mystery of the Mind: A Critical Study of Consciousness and the Human Brain. Princeton, Princeton University Press (1975)
5. Goldberg, M.E.: The control of gaze. In: Kandel, E.R., Schwartz, J.H., Jessel, J.H. (eds.) Principles of Neuroscience, 4th edn, pp. 782–800. McGraw Hill, New York (2000)
6. McNarry, A.F., Goldhill, D.R.: Simple bedside assessment of level of consciousness: comparison of two simple assessment scales with the Glasgow Coma scale. Anaesthesia 59, 34–37 (2004)
7. Parvizi, J., Damasio, A.R.: Neuroanatomical correlates of brainstem coma. Brain 126, 1524–1536 (2003)
8. Moruzzi, G., Magoun, H.W.: Brain stem reticular formation and activation of the EEG. Electroencephalogr. Clin. Neurophysiol. 1, 455–473 (1949)
9. Magoun, H.W.: An ascending reticular activating system in the brain stem. Arch. Neurol. Psychiatry 67(2), 145–154 (1952)
10. Batini, C., Moruzzi, G., Palestino, M., Rossi, G.F., Zanchetti, A.: Persistent patterns of wakefulness in the pretrigeminal midpontine preparation. Science 128, 30 (1958)
11. Mai, J., Paxinos, G.: The Human Nervous System. Elsevier, Amsterdam (2011)
12. Kandel, E.R., Schwartz, J.H., Jessell, T.M., Siegelbaum, S.A., Hudspeth, A.J.: Principles of Neural Science. McGraw-Hill, New York (2012)
13. Posner, J.B., Clifford, B.S., Schiff, N.D., Plum, F.: Plum and Posner's Diagnosis of Stupor and Coma, 4th edn. Oxford University Press, New York (2007)
14. Laureys, S., Gosseries, O., Tononi, G. (eds.): The Neurology of Consciousness: Cognitive Science and Neuropathology, 2nd edn. Waltham, MA, New York (2016)
15. Bloom, F.E.: What is the role of general activating systems in cortical function? In: Rakic, P., Singer, W. (eds.) Neurobiology of Neocortex, pp. 407–421. Wiley, New York (1997)
16. Hirata, A., Castro-Alamancos, M.A.: Neocortex network activation and deactivation states controlled by the thalamus. J. Neurophysiol. 103, 1147–1157 (2010)
17. Damasio, A.R., Grabowski, T.J., Bechara, A., Damasio, H., Ponto, L.L., Parvizi, J., Hichwa, R.D.: Distinctive patterns of subcortical and cortical brain activation associated with self-generated emotions and feelings. Nat. Neurosci. 3(10), 1049–1056 (2000)
18. Llinas, R., Ribury, U., Contreras, D., Pedroarena, C.: The neuronal basis of consciousness. Philos. Trans. R. Soc. Lond. B Biol. Sci. 353, 1841–1849 (1998)
19. Dehaene, S., Changeux, J.P.: Neural mechanisms for access to consciousness. In: Gazanniga, M.S. (ed.) The Cognitive Neurosciences, pp. 1145–1147. Norton, New York (2004)
20. Crick, F.: Function of the thalamic reticular complex: The searchlight hypothesis. Proc. Natl. Acad. Sci. U.S.A. 81, 4586–4590 (1984)
21. Koyama, T., McHaffie, J.G., Luarienti, P.J., Coghill, R.C.: The subjective experience of pain: Where expectations become reality. Proc. Natl. Acad. Sci. U.S.A. 102, 12950–12955 (2005)
22. Steriade, M.: Neuromodulatory Systems of Thalamus and Neocortex. Semi. Neurosci. 7(5), 361–370 (1995)
23. Hirata, A., Castro-Alamancos, M.A.: Neocortex network activation and deactivation states controlled by the thalamus. J. Neurophysiol. 103, 1147–1157 (2010)
24. Harris, K.D., Thiele, A.: Cortical state and attention. Nat. Rev. Neurosci. 12, 509–523 (2011)
25. Maldonato, N.M.: The Ascending Reticular Activating System. The Common Root of Consciousness and Attention, vol. 26, pp. 333–344. Springer, Berlin (2014)
26. Rigas, P., Castro-Alamancos, M.A.: Thalamocortical up states: differential effects of intrinsic and extrinsic cortical inputs on persistent activity. J. Neurosci. 27, 4261–4272 (2007)
27. Penfield, W.: Lights in the great darkness. Harvey cushing oration. J. Neurosurg. 35, 377–383 (1971)
28. Penfield, W.: Some observations on the cerebral cortex of man. In: Proceedings of the Royal Society of London. Series B, Biological Sciences, vol. 134, pp. 329–47 (1947)
29. Yeo, S.S., Chang, P.H., Jang, S.H.: The ascending reticular activating system from pontine reticular formation to the thalamus in the human brain. Front Hum Neurosci 7, 416 (2013)
30. Maldonato, N.M., Dell'Orco, S., Springer, M.: Rethinking consciousness: some hypothesis on the role of the ascending reticular activating system in global workspace. In: Apolloni, B. (ed.) Neural Nets WIRN11. IOS Press, Amsterdam (2011)

31. Damasio, A.R., Grabowski, T.J., Bechara, A., Damasio, H., Ponto, L.L., Parvizi, J., Hichwa, R.D.: Distinctive patterns of subcortical and cortical brain activation associated with self-generated emotions and feelings. Nat. Neurosci. **3**(10), 1049–1056 (2000)
32. Oliverio, A., Maldonato, N.M.: The creative brain. In: CogInfoCom, 5th IEEE International Conference on Cognitive Infocommunications, November 5–7, pp. 527–532 (2014)
33. Maldonato, N.M.: From Neuron to Consciousness: For an Experience-Based Neuroscience. World Futures **65**(2), 80–93 (2009)
34. Kocsis, B., Varga, V., Dahan, L., Sik, A.: Serotonergic neuron diversity: identification of raphe neurons with discharges time-locked to the hippocampal theta rhythm. Proc. Natl. Acad. Sci. U.S.A. **103**, 1059–1064 (2006)
35. Grzanna, R., Molliver, M.E.: Cytoarchitecture and dendritic morphology of central noradrenergic neurons. In: Hobson, J.A., Brazier, M.A.B. (eds.) The Reticular Formation Revisited: Specifying Function for a Nonspecific System, pp. 83–97. Raven Press, New York (1980)
36. Monti, J.M., Jantos, H.C.: The roles of dopamine and serotonin, and of their receptors, in regulating sleep and waking. Prog. Brain Res. **172**, 625–646 (2008)
37. Parvizi, J., Damasio, A.: Consciousness and the brainstem. Cognition **79**, 135–160 (2001)
38. Maldonato, N.M., Oliverio, A., Esposito, A.: Neuronal symphonies: musical improvisation and the centrencephalic space of functional integration. World Futures **73**, 1–20 (2017)
39. Damasio, A., Carvalho, G.B.: The nature of feelings: Evolutionary and neurobiological origins. Nat. Rev. Neurosci. **14**(2), 143–152 (2013)
40. Zeki, S., Bartels, A.: The asynchrony of consciousness. Proc. Roy. Soc. London **265**, 1583–1585 (1998)
41. Libet, B.: Mind Time. Harvard University Press, Cambridge Mass (2004)
42. Haggard, P., Clark, S., Kalogeras, J.: Voluntary action and conscious awareness. Nat. Neurosci. **5**(4), 382–385 (2002)
43. Berlucchi, G., Aglioti, S.M.: The body in the brain. Exp. Brain Res. **200**, 25–35 (2010)
44. Ey, H.: Conscioussnes: A Phenomenological Study of Being Conscious and Becoming Conscious. Indiana University Press, Bloomington (1978)
45. Changeux, J.P.: Conscious processing: implications for general anesthesia. Curr Opin Anaesthesiol. **25**, 397–404 (2012)
46. Mashour, G.A., Alkire, M.T.: Evolution of consciousness: phylogeny, ontogeny, and emergence from general anesthesia. Proc. Natl. Acad. Sci. U.S.A. **110**(Suppl. 2), 10357–10364 (2013)
47. Posner, J.B., Clifford, B.S., Schiff, N.D., Plum, F.: Plum and Posner's Diagnosis of Stupor and Coma, 4th edn. Oxford University Press, New York (2007)
48. Sanders, R.D., Tononi, G., Laureys, S., Sleigh, J.: Unresponsiveness not equal Unconsciousness. Anesthesiology **116**, 946–959 (2012)
49. Långsjö, J.W., Alkire, M.T., Kaskinoro, K., Hayama, H., Maksimow, A., Kaisti, K.K., Aalto, S., Aantaa, R., Jääskeläinen, S.K., Revonsuo, A., Scheinin, H.: Returning from Oblivion: Imaging the neural core of consciousness. J. Neurosci. **32**(14), 4935–4943 (2012)
50. Guldenmund, P., Vanhaudenhuyse, A., Boly, M., Laureys, S., Soddu, A.: A default mode of brain function in altered states of consciousness. Arch. Ital. Biol. **150**(2–3), 107–121 (2012)
51. Boly, M., Phillips, C., Tshibanda, L., Vanhaudenhuyse, A., Schabus, M., Dang-Vu, T.T., et al.: Intrinsic brain activity in altered states of consciousness: how conscious is the default mode of brain function? Ann NY Acad Sci **1129**, 119–129 (2008)
52. Maldonato, N.M., Anzoise, A.: Homo-Machina Visual Metaphors. Representations of Consciousness and Scientific thinking, pp. 437–451. Springer, Berlin, Heidelberg (2013)
53. Maldonato, N.M., Valerio, P.: Artificial entities or moral agents? How AI is changing human evolution. In: Esposito, A., Faundez-Zanuy, M., Morabito, F.C., Pasero, E. (eds.) Multidisciplinary Approaches to Neural Computing. Series: Smart Innovation, Systems and Technologies, vol. 69. Springer, Berlin (2017)

Chapter 13
Does Neuroeconomics Really Need the Brain?

Nelson Mauro Maldonato, Luigi Maria Sicca, Antonietta M. Esposito and Raffaele Sperandeo

Abstract The systematic study of biological basis of behavior and of the process involved in economical choices has outlined a new paradigm of research: neuroeconomics. Now the intersection between neuroscience, psychology and economics, neuroeconomics presents itself as an alternative to the neoclassical vision on economics, according to which the homo oeconomicus acts within the bonds of a formalizing rationality tending to the maximization of the anticipated utility. Brain imagining methods have shown that the decision-making processes activate the frontal lobe and the limbic system above all, a big circonvolution running through the callous body on the medial surface of the hemispheres, extending itself down, responsible for the regulation of emotional phenomena. Reinforcing such a tendency, we find the injury paradigm. It was observed that frontal lobe injuries harm the capacity of making advantageous decisions either in one's own behalf or in others, as well as decisions according to the social conventions. In this paper, we will try to show that if, by the one hand, the neuro visual methods have given us a great amount of data, on the other hand, using them uncritically, with the recurrent confusion between "correlation" and "causal relation"—contemporary microevents indicate only simple correlations, and no cause-effect relation—risks to stress the relevant explanatory gap regarding the abstract ideal of understanding the nature of the brain.

N. M. Maldonato (✉)
Dipartimento di Neuroscienze e Scienze Riproduttive ed Odontostomatologiche, Università di Napoli Federico II, Naples, Italy
e-mail: nelsonmauro.maldonato@unina.it

L. M. Sicca
Dipartimento di Economia, Management, Istituzioni, Università degli Studi di Napoli Federico II, Naples, Italy
e-mail: lumsicca@unina.it

A. M. Esposito
Istituto nazionale di Geofisica e Vulcanologia, Sezione di Napoli, Osservatorio Vesuviano, Naples, Italy
e-mail: antonietta.esposito@ingv.it

R. Sperandeo
Sipgi, Scuola di Psicoterapia Gestaltica Integrata, Torre Annunziata, Italy
e-mail: raffaele.sperandeo@gmail.com

© Springer International Publishing AG, part of Springer Nature 2019
A. Esposito et al. (eds.), *Quantifying and Processing Biomedical and Behavioral Signals*, Smart Innovation, Systems and Technologies 103,
https://doi.org/10.1007/978-3-319-95095-2_13

Keywords Decision making · Neuroeconomics · Brain imaging · Somatic marker · Behavioral economics

13.1   From Behavioral Economics to Neuroeconomics

For our species, decisions has always represented the strongest factor of selection and adaptation, of transformation and creativity [1]. Through its long evolutionary course, the human brain has molded itself in order to fulfill its own biological and psychological needs, continually drawing information from the external environment and elaborating quick and effective answers to it [2]. It was through quick, not necessarily conscious choices that our ancestors survived extremely complex and often dangerous contexts. At least in the early stages of human evolution, strategies of formal reasoning would have been improper in order to operate choices in restricted time and under conditions of uncertainty. It is much more 'economical' to take advantage of certain signals instead of analyzing the situation in every detail [3].

Today, as in the past, when under risky and uncertain conditions, when facing decisions that concern single events, the only possibility of analyzing every possibility is relying on one's own emotional devices and on pre-selected cognitive schemes—heuristic ones—informal, logical inferences which allow the decision-maker to reach quick conclusions without too much cognitive effort [4]. Beyond an abstract physical world, the environment in which individuals interact is constituted, above all, by our counterparts and by other species occupying our ecological niche. The stability of the environment is a key element in trying to foresee future occurrences and is also a function involving decisions, choices and events over which we hardly ever have any control, as with acts of will, which give us but the illusion of making decisions [5].

In neoclassical economics, the identity of the homo oeconomicus is associated with maximizing and its subjective unity is indifferent to the others and increases according to the higher number of goods one has, even if it grows shorter in the margins: therefore, the more equal the goods, the more useful they are, but this is not directly proportional [6]. He acts according to criteria of rationality connected to the principles of the expected utility hypothesis. Kahneman and Tversky proved [7], however, that decision-making under uncertainty doesn't follow the rules of probability in the economic theory, but schemes and propensities (which often denote proper cognitive styles [8]) selected according to the capacity of better adapting oneself to the various environmental contexts [9]. The heuristics are more than simple thinking shortcuts: they are sophisticated expressions of selective pressures that had had place in social contexts of great complexity. The concept of homo oeconomicus is, on the other hand, a rather peculiar form of cultural evolution. Neoclassical economics has considered him capable of rational choices and of foreseeing human behavior: but its almost total lack of ability in developing predictive analysis and reproducible experiments turns economics into something quite different from a science [10].

13.2   The Dark Side of Emotions: The Indifference Towards the Future Due to Orbitofrontal Damage

In the last decade, the psychology of decision-making is getting progressively close to knowing the neuronal basis involved in decisional behavior [11]. A new discipline, defined as neuroeconomics, has actually opened a new season of research through "functional" experiments made possible by brain imaging methods invasive to a very low degree [12]. The individuation of physiological equivalents—called "somatic markers"—makes it possible now to tell the difference between various kinds of behavior and to make the nature of decisions and their psychobiological implications clear. Analyzing "somatic markers" [13] is an efficient key to understand the paradoxical behavior of patients who suffered injuries in the orbitofrontal cortex. When this area is damaged the representations that guide and produce an action enter the working memory without their emotional content [14]. Such a patient can still reason without problems, but in a detached, almost indifferent way. For instance, the death of somebody one loves is not followed by the experience of pain often associated to similar situations. Orbitofrontal damage eliminates the emotional elaboration of affective memories, provoking shifts in skin conductance responses (S. C. R.) mediated by the autonomous nervous system [15]. In patients who had orbitofrontal cortex injuries, the register of physiological indicators results totally flat: they make inappropriate choices and are incapable of generating anticipated skin conductance responses in relation to choices that, in every case, remain inadequate. For example, supposedly alarming situations provoke no reaction in the patient. That shows emotional flatness and emotional physiological indicator are directly proportional in these patients [16].

These data lead us to ask: the physiological and emotional answers really mediate the decision-making process? One has tried to answer to this question through the risk taking test, in which certain stimuli are associates to (finite) rewards and financial sanctions [17]. The individuals, indeed, are free to choose their cards from two decks, learning through trial and error which of both allows one to score higher. The aim of the game is getting the higher amount of money possible by choosing a deck at a time and looking at its first card. The cards in certain decks can lead to high sums (U$100), but can also hide grave penalties (up to U$1250); in other, rewarding decks, the penalties are softer (near U$50 and no more than U$100). It happens, then, that the controlled individuals gradually choose the less risky decks while patients who had suffered orbitofrontal injuries prefer the more risky ones, maybe because they are attracted by the recurring hundred-dollar winnings, despite the very high penalties. The most interesting aspect in both groups is in S. C. R. Looking at a card both groups of individuals show a transitory S. C. R. increase, therefore an answer from the autonomous nervous system to the rewards or to the penalties. Within time, however, in controlled individuals those shifts have a precocious onset. That is to say while these individuals prepare themselves to choose the card from the most risky deck, their S. C. R. raises intensely. On the contrary, the S. C. R. of patients who suffered orbitofrontal injury don't who any change in the same situation and, therefore, there is no physiological evidences of the fact that their decision is mediated by emotion [18, 19].

What is not clear yet is the nature of the relation between the frontal and the limbic areas [20, 21]. One thinks of the aversion to ambiguity. There are people who, in face of the high chance of winning, prefer to bet on clear events rather than on vague ones. On the other hand, while they face a high chance of loosing, they prefer vagueness to clarity. Here neuroimaging experiments show the activations of a zone in the limbic system and of a part of cerebral cortex as if both zones were 'talking' to each other. Actually, the interpretation of these events is highly controversial [22]. If, from one hand, the aversion to ambiguity could be generated from fear; on the other hand, ambiguity could represent a cognitive process of data elaboration as others, based on shorter availability of information [23]. The consequence of both positions change the very predictability of behaviors [24]. In reality, is we claim that the reaction of the brain to ambiguity is emotionally based, it is difficult to believe it could learn how to deal with it. If, however, we consider it as cognitively based, then it means one can slowly get used to ambiguity [25]. In every case, fMRI images don't help us to take the necessary conclusions from such complex behavior [26].

13.3 Reverse Inference and the Principle of Indetermination in the Human Brain

In every likelihood, neuroeconomics, as other spheres of research, is exposed to relevant problems in which concerns both method and epistemology [27]. If one considers, for example, the question of overlapping of data and methodology between economic models and psychological theories [28], it is clear that the central role assigned to "measurability" inevitably induces neuroeconomical research to reconsidering traditional economic models, according to which all the choices must be traced back (at least ideally) to quantifiable factors such as prices, quantities and probabilities [29]. It all leads us to inquire: in what sense the discovery of new brain patterns can guide the creation of new economical hypotheses? Moreover: if a brain scan is a calculation of probabilities of what will occur in a specific brain area, who can predict for sure what happens in a given circumstance? The only predictable thing is the probability of diverse events. There is an indetermination principle concerning even the human brain [30, 31].

But we shall look closely at the possible meaning of the fMRI methods used nowadays. As it is known, the technique they use is predominantly BOLD (Blood Oxygenation Level Dependent) to study the flow of hematic volume in certain brain areas [32]. It is about techniques to study cortical perfusion. As in vitro studies have clarified, the activation of a cortical region produces an alteration in microcirculation, determining a local increment of the two mentioned parameters [33], observable thanks to an increment in the quota of oxygenated hemoglobin in comparison with deoxygenated hemoglobin [34]. Therefore, a higher consummation of oxygen is not directly observed, but the increment in the percentage of oxygenated hemoglobin furnished by microcirculation to an activated cortical zone. This phenomenon assumes, therefore, that brain activity reflects the metabolic engagement of a cortical area: then, this is not observed directly, but through the hemodynamic responses related to it [35]. In other words, we only know that neural activity produces activation patterns in microcirculation, in its relative times of activation [36, 37]. The same must be said about PET and SPECT, which measure the efficiency of metabolic ways through radioactive decay [38]. Same as EEG and MEG [39], which measure the effect produced by synchronic activity of neuronal population in the variable electromagnetic field (on the superficial dermis of the skull). How valid they are is what join these methods [40]. It is about, however, phenomena whose systematic correlation with certain events are demonstrable at cognitive, behavioral levels, as well as in the level of introspective reports. Hence, describing a casual relation or a sequence of casual relations on this basis is implausible [41, 43]. Naturally, that doesn't mean those correlations are not properly scientific; indeed, its models/patterns can be validated (or falsified) and, doubtless, allow predictions [43, 44]. However, without building usual chains, they describe, they don't explain.

13.4   Conclusions

The idea according to which new methods of brain imaging can provide us with meaningful answers or hints about decisional behavior goes beyond the goal of a model that would overcome the classical model, the homo oeconomicus. Despite the excitement and the promises, the new methods of studying the brain don't give us evidence to allow us to overcome the historical handicap of classical economics: that is to say, describing rather than explaining, rationalizing ex post about past behavior rather than predicting future behavior [45]. This methodological and instrumental apparatus shows limits that are still relevant to the complexity of the very systems it tries to explain: they seem, however, to make predictions as a science that defines itself as a science should do. A statistic-or probability-related approach to decisional behavior is not enough on its own: given the extent of the calculations involved, it is even useless. This would not only generate a bulk of computational work out of the range of the human brain, but also statistic estimates on the probability of single, individual mathematical meaningless future behavior.

References

1. Maldonato, N.M., Dell'Orco, S.: How to make decisions in an uncertain world: heuristics, biases, and risk perception. World Futures **67**(8), 569–577 (2011)
2. Maldonato, N.M., Dell'Orco, S.: Toward an evolutionary theory of rationality. World Futures **66**(2), 103–123 (2010)

3. Maldonato, N.M., Dell'Orco, S.: Making Decision Under Uncertainty: Emotions, Risk and Biases. Smart Innovation, Systems and Technologies, vol. 37, pp. 293–302. Springer, Berlin (2015)
4. Gigerenzer, G., Selten, R.: Bounded Rationality: The Adaptive Toolbox. MIT Press, Cambridge, MA (2002)
5. Gazzaniga, M.S.: Who's in Charge? Free Will and the Science of the Brain. HarperCollins, New York (2011)
6. Neumann, J., Morgenstern, O.: Theory of Games and Economic Behavior. Princeton University Press, Princeton (1947)
7. Tversky, A., Kahneman, D.: The framing of decisions and the psychology of choice. Science **211**(448), 453–458 (1981)
8. Scott, S., Bruce, R.: Decision-making style: the development of a new measure. Educ. Psychol. Measur. **5**, 818–831 (1995)
9. Tversky, A., Sattah, S., Slovic, P.: Contingent weighting in judgment and choice. Psychol. Rev. **95**(3), 371–384 (1988)
10. Maldonato, N.M., Dell'Orco, S.: The predictive brain. World Futures **68**(6), 381–389 (2012)
11. Camerer, C., Loewenstein, G., Prelec, D.: Neuroeconomics: how neuroscience can inform economics. J. Econ. Lit. **43**, 9–64 (2005)
12. Ogawa, S., Tank, D.W., Menon, R., et al.: Intrinsic signal changes accompanying sensory stimulation: functional brain mapping with magnetic resonance imaging. Proc. Natl. Acad. Sci. U.S.A. **89**, 5951–5955 (1992)
13. Dunn, B.D., Dalgleish, T., Lawrence, A.D.: The somatic marker hypothesis: a critical evaluation. Neurosci. Biobehav. Rev. **30**, 239–271 (2006)
14. Damasio, A.R.: The somatic marker hypothesis and the possible functions of the prefrontal cortex. Philos. Trans. R. Soc. Lond. **351**(1346), 1413–1420 (1996)
15. Damasio, A.R., Tranel, D., Damasio, H.: Individuals with sociopathic behavior caused by frontal damage fail to respond autonomically to social stimuli. Behav. Brain Res. **41**(2), 81–94 (1990)
16. Damasio, A.R.: The frontal lobes. In: Heilman, K.M., Valenstein, E. (eds.) Clinical Neuropsychology. Oxford University Press, New York (1979)
17. Schmitt, W.A., Brinkley, C.A., Newman, J.P.: Testing Damasio's somatic marker hypothesis with psychopathic individuals: risk takers or risk averse? J. Abnorm. Psychol. **108**(3), 538–543 (1999)
18. Maldonato, N.M.: Undecidable decisions: rationality limits and decision-making heuristics. World Futures **63**(1), 28–37 (2007)
19. Bechara, A., Damasio, H.: Decision-making and addiction (part I): impaired activation of somatic states in substance dependent individuals when pondering decisions with negative future consequences. Neuropsychologia **40**, 1675–1689 (2002)
20. Fix, J.D.: Theory of Games and Economic Behavior. Neuroanatomy. Lippincott Williams & Wilkins, Philadelphia (2008)
21. Finger, S.: Defining and Controlling the Circuits of Emotion. Origins of Neuroscience: A History of Explorations into Brain Function. Oxford University Press, Oxford (2001)
22. Al-Najjar, N.I., Weinstein, J.: The ambiguity aversion literature: a critical assessment. Econ. Philos. **25**, 249–284 (2009)
23. Maldonato, N.M., Dell'Orco, S., Sperandeo, R.: When intuitive decisions making, based on expertise, may deliver better results than a rational, deliberate approach. In: Esposito, A., Faundez-Zanuy, M., Morabito, F.C., Pasero, E. (eds.) Multidisciplinary Approaches to Neural Computing. Springer, Cham (2018)
24. Poline, J.B., Worsley, K.J., Evans, A.C., Friston, K.J.: Combining spatial extent and peak intensity to test for activations in functional imaging. NeuroImage **5**(2), 83–96 (1997)
25. Ramnani, N., Owen, A.M.: Anterior prefrontal cortex: insights into function from anatomy and neuroimaging. Nat. Rev. Neurosci. **5**, 184–194 (2004)
26. Glimcher, P.W., Camerer, C., Poldrack, R.A., Fehr, E.: Neuroeconomics: Decision Making and the Brain. Academic Press, Cambridge (2008)

27. Schonberg, T., Bakkour, A., Hover, A.M., Mumford, J.A., Nagar, L., Perez, J., Poldrack, R.A.: Changing value through cued approach: an automatic mechanism of behavior change. Nat. Neurosci. **17**(4), 625–630 (2014)
28. Maldonato, N.M., Dell'Orco, S.: The natural logic of action. World Futures **69**(3), 174–183 (2013)
29. Davis, T., LaRocque, K.F., Mumford, J.A., Norman, K.A., Wagner, A.D., Poldrack, R.A.: What do differences between multi-voxel and univariate analysis mean? How subject-, voxel-, and trial-level variance impact fMRI analysis. Neuroimage **97**, 271–283 (2014)
30. Park, M., Koyejo, O., Ghosh, J., Poldrack, R., Pillow, J.: Bayesian structure learning for functional neuroimaging. In: Proceedings of the Sixteenth International Conference on Artificial Intelligence and Statistics, pp. 489–497 (2013)
31. Koyejo, O., Patel, P., Ghosh, J., Poldrack, R.A.: Learning predictive cognitive structure from fMRI using supervised topic models. In: International Workshop on Pattern Recognition in Neuroimaging (PRNI), pp. 9–12 (2013)
32. Friston, K., Rigoli, F., Ognibene, D., Mathys, C., Fitzgerald, T., Pezzulo, G.: Active inference and epistemic value. Cogn. Neurosci. **6**(4), 187–214 (2015)
33. Ogawa, S., Lee, T.M., Nayak, A.S., Glynn, P.: Oxygenation-sensitive contrast in magnetic resonance image of rodent brain at high magnetic fields. Mag. Res. Med. **14**, 68–78 (1990)
34. Jezzard, P., Matthews, P.M., Smith, S.M.: Functional Magnetic Resonance Imaging. An Introduction to Methods. Oxford University Press, Oxford (2003)
35. Raichle, M.E.: Behind the scenes of functional brain imaging: a historical and physiological perspective. Proc. Natl. Acad. Sci. U.S.A. **95**(3), 765–772 (1998)
36. Friston, K.: Functional integration and inference in the brain. Prog. Neurobiol. **68**(2), 113–143 (2002)
37. Friston, K., Kahan, J., Razi, A., Stephan, K.E., Sporns, O.: On nodes and modes in resting state fMRI. Neuroimage **99**, 533–547 (2014)
38. Forsting, M., Jansen, O.: MR Neuroimaging: Brain, Spine, Peripheral Nerves. Thieme, New York (2017)
39. Riitta, H., Puce, A.: MEG-EEG Primer. Oxford University Press, Oxford (2017)
40. Friston, K.: The history of the future of the Bayesian brain. Neuroimage **62–248**(2), 1230–1233 (2012)
41. Friston, K.: A theory of cortical responses. Philos. Trans. R. Soc. Lond. B Biol. Sci. **360**, 815–836 (2005)
42. Friston, K.: The free-energy principle: a unified brain theory? Nat. Rev. Neurosci. **11**(2), 127–138 (2010)
43. Brakewood, B., Poldrack, R.A.: The ethics of secondary data analysis: considering the application of Belmont principles to the sharing of neuroimaging data. Neuroimage **82**, 671–676 (2013)
44. Davis, T., Poldrack, R.A.: Measuring neural representations with fMRI: practices and pitfalls. Ann. N. Y. Acad. Sci. **1296**, 108–134 (2013)
45. Maldonato, N.M., Dell'Orco, S.: Adaptive and evolutive algorithms: a natural logic for artificial mind. In: Esposito, A., Jain, L.C. (eds.) Toward Robotic Socially Believable Behaving Systems, vol. II. Springer, Cham (2016)

Chapter 14
Coherence-Based Complex Network Analysis of Absence Seizure EEG Signals

Nadia Mammone, Cosimo Ieracitano, Jonas Duun-Henriksen, Troels
Wesenberg Kjaer and Francesco Carlo Morabito

Abstract This paper addressed the issue of epileptic absence seizures developing
a complex brain network model based on the estimation of the coherence between
electroencephalographic (EEG) signals. The EEG signals indeed reflect the abnormalities in the cortical electrical activity caused by epilepsy. A dataset of 10 absence
patients was analyzed, including 63 seizures. The model was analyzed over the time to
assess if changes in the network parameters matched the brain state (ictal: seizure),
(non-ictal: seizure free). During the ictal states, the characteristic path length (λ)
decreased and the global efficiency (GE), the average clustering coefficient (CC) and
the small worldness (SW) increased, as expected, because of the abnormal synchronization associated with absence seizure onset. The connection matrices preceding
the ictal states (8 s before) were thresholded and the corresponding connectivity scalp
maps, showing the active links between EEG channels, were displayed. Such connectivity maps showed the interaction between channels and provided information
about the abnormal recruitment mechanism associated with seizure development: the
involvement of the cortical areas appears progressive and that every subject exhibited
peculiar recurrent patterns of area activation.

N. Mammone (✉)
IRCCS Centro Neurolesi Bonino-Pulejo, Via Palermo c/da Casazza, SS. 113, Messina, Italy
e-mail: nadia.mammone@irccsme.it

C. Ieracitano · F. C. Morabito
Mediterranea University of Reggio Calabria, Via Graziella,
Feo di Vito, 89060 Reggio Calabria, Italy
e-mail: cosimo.ieracitano@unirc.it

F. C. Morabito
e-mail: morabito@unirc.it

J. Duun-Henriksen
UNEEG™ Medical A/S, Nymollevej 6, DK-3540, Lynge, Denmark
e-mail: jh@hyposafe.com

T. W. Kjaer
Department of Neurology, Center of Neurophysiology,
Zealand University Hospital, Roskilde, Denmark
e-mail: neurology@dadlnet.dk

© Springer International Publishing AG, part of Springer Nature 2019
A. Esposito et al. (eds.), *Quantifying and Processing Biomedical and Behavioral
Signals*, Smart Innovation, Systems and Technologies 103,
https://doi.org/10.1007/978-3-319-95095-2_14

143

Keywords Complex network analysis · Coherence · Electroencephalogram
Childhood absence epilepsy

14.1 Introduction

Epilepsy is a neurological disease that causes recurrent seizures and affects about
1% of the world population. In the last decades, worldwide researchers have devoted
considerable efforts to develop algorithms, based on electroencephalographic (EEG)
signals processing, for the automatic detection and early detection of the epileptic seizures, nonetheless, seizures are still unpredictable in practice. In this work,
the issue of detecting seizures in Childhood Absence Epilepsy (CAE) patients is
addressed. CAE is an idiopathic generalized kind of epilepsy [1] characterized by
recurrent episodes of transient awareness' impairment called "absence seizures".
Subjects who experience seizures must undergo EEG examination, which is com-
fortable, safe and non-invasive. In order to record the scalp potentials produced by
the cortical electrical activity, a set of EEG electrodes is uniformly placed on the
scalp, usually by means of a cap, in order to cover uniformly all the areas of the
brain (frontal, temporal, occipital, central, parietal). The EEG acquisition system is
connected to a computer so that the EEG traces are digitally stored. Many methodologies have been proposed so far to analyze the EEGs of CAE patients. In 2002
a new symbolic complexity measure, named Permutation Entropy (PE), was introduced by Bandt and Pompe [2], Cao et al. [3] applied PE to classify different stages
of epileptic activity in the intracranial EEG signals (iEEG) recorded from three
intractable patients. Bruzzo et al. [4] investigated the vigilance changes in epileptic
patients through the quantification of EEG complexity. Ouyang et al. used PE and
multiscale permutation entropy (MPE) [5] to analyze human EEG signals at different
absence seizure states: Linear Discriminant Analysis (LDA), when based on MPE,
detected absence seizures with a 90.6% sensitivity. Zhu et al. [6] used PE as feature for K-means, Multi-Scale K-means (MSK-means) and support vector machine
(SVM) based classification, with the purpose of detecting seizures and localizing the
epileptogenic zone. In order to have an overall view of PE levels distribution over
the scalp in absence patients, its topography was analyzed. PE topography showed
that the frontal temporal lobes of CAE patients exhibited relatively high PE levels, as compared to the parieto-occipital areas [7, 8]. Summarizing, PE, PE-derived
descriptors [9], and other entropy descriptors [10] have been extensively used in
the analysis of absence EEG, however, they are univariate descriptors and therefore
EEG signals are processed independently, missing the interactions between channels. Since the goal of the present work is to study the epileptic brain as a complex
system, the quantification of the interaction between EEG channels is necessary. In
fact, the channels cover the different cortical areas, which are the elements of the
complex system under analysis. This is the reason why the attention was focused on
bivariate descriptors of EEG activity. By measuring the interaction between every
pair of EEG signals, the interaction between every pair of cortical areas can be indi-

rectly estimated and a complex network model of interaction can be constructed. In particular, Coherence was considered in the present work due to its ability to detect the synchronization changes triggered by absence seizures [11, 12]. Rotondi et al. [12] used Partial Directed Coherence (PDC) to measure the interactions between the electrodes during the inter-ictal EEG segments in CAE patients. Abnormal cortical network activity was detected in the inter-ictal state in alpha band (8–13 Hz). Ravish et al. [13] proposed the joint use of wavelet transform, coherence and phase synchrony to classify ictal vs inter-ictal EEG epochs. Rodrigues et al. [14] developed a PDC-based graph model of EEGs of CAE patients to classify the channels according to their source/sink nature.

The present has a double goal: (1) developing a coherence-based complex network model in order to assess if the onset of epileptiform activity affects the features of the model; (2) finding possible patterns of interaction between channels in order to find out how channels are recruited during the evolution towards and during the ictal events. The proposed approach differs from the related studies in the literature. First of all, in previous studies focused on the complex networks analysis of absence seizure EEG, only one seizure per patient was considered [11] whereas, in the present study, all the available recorded seizures were analysed (63 in total) in order to look for possible recurrent patterns that could be associated with seizure development. Here, the EEG is partitioned into overlapping epochs and then processed epoch by epoch. Secondly, artifactual epochs were not rejected in order to follow the EEG dynamics continuously and smoothly. Furthermore, to the authors' best knowledge, this is the first time that patterns of interaction between channels have been investigated to find possible relationships between channels recruitment and seizure development. Since the results achieved over the analyzed dataset of 10 CAE patients look very promising, the dataset and the analysis will be extended in the near future. The paper is organized as follows: in Sect. 14.2, the analysed EEG dataset and the coherence based complex network model are described. Section 14.3 illustrates the results of the complex network feature estimation as well as the analysis of patterns of electrode recruitment associated with seizure onset. Finally, Sect. 14.4 draws the conclusions.

14.2 Coherence-Based EEG Complex Network Analysis

14.2.1 EEG Recording and Pre-processing

The analysed EEG data were provided by $UNEEG^{TM}$ medical A/S (Lynge, Denmark) within a research cooperation agreement between NeuroLab laboratory and $UNEEG^{TM}$ medical A/S. Nineteen EEG electrodes were placed over the scalp according to the international 10/20 system. The EEG acquisition systems used in this work are *Stellate Harmonie* (Stellate Systems, Inc., Montreal, Quebec, Canada) and *Cadwell Easy II* (Cadwell Laboratories, Inc., Kennewick, WA). All the paroxysms were manually identified in the EEG recordings by an expert, board-certified, epileptologist. The dataset consists in 10 EEGs recorded from CAE patients (mean age = 7.44 years), with an average length of 24.5 min, including 63 seizures in total. The *19*-channels EEG is recorded, band-pass filtered (between 0.5 and 32 Hz, because absence seizure activity was shown to be mainly in this range [15]), sampled at 200 Hz and digitally stored. The EEG is then processed offline: it is segmented into M overlapping epochs (the width of each epoch is $2s$ and the overlap is $1s$) and then it is processed epoch by epoch. The width was set at $2s$ in order to have epochs short enough to capture the short, transient events, whereas the overlap ensures that the time series of the estimated descriptor is smooth. For every k-th epoch, the corresponding EEG segment $EEG(k)$ (with $k = 1, \ldots, M$), is processed and the coherence C_{x_i,x_j} between every pair x_i and x_j is estimated. In the proposed complex network model, the nodes of the graph represent the EEG electrodes (each one associated with the corresponding cortical sub-area) and the weight between nodes x_i and x_j is the coherence C_{x_i,x_j} between the EEG signals recorded at electrodes x_i and x_j. All the algorithms were implemented in MATLAB R2016a (The MathWorks, Inc., Natick, MA, USA).

14.2.2 Coherence Estimation

Given two signals x_i and x_j, the magnitude squared coherence between them is defined as:

$$C_{x_i,x_j}(f) = \frac{|P_{x_i,x_j}(f)|^2}{P_{x_i,x_i}(f)P_{x_j,x_j}(f)}$$

where f is the frequency, $P_{x_i,x_i}(f)$ and $P_{x_j,x_j}(f)$ are the Power Spectral Densities (PSD) of x_i and x_j, respectively, and $P_{x_i,x_j}(f)$ is the cross power spectral density between x_i and x_j.

Coherence C_{x_i,x_j} is a measure of similarity between x_i and x_j (it is also bounded, in fact $0 \leq C_{x_i,x_j} \leq 1$). In this paper, coherence was estimated using Welch's averaged, modified periodogram algorithm [16]. Given the k-th EEG epoch under analysis, coherence $C_{x_i,x_j}^k(f)$ is estimated for every frequency f, then the values $C_{x_i,x_j}^k(f)$ are averaged within the frequency range under consideration $f_L - f_U$:

$$\overline{C}_{x_i,x_j}^k = \frac{1}{f_U - f_L} \int_{f_L}^{f_U} C_{x_i,x_j}^k(f)df$$

where $f_L = 0.5\,\text{Hz}$ and $f_U = 32\,\text{Hz}$. In conclusion, for every analyzed epoch $EEG(k)$ and for every pair of electrodes (x_i, x_j), an average value of coherence $\overline{C}_{x_i,x_j}^k$ is estimated and it becomes the i, j element of the connection matrix $\mathbf{CM}^k$ which describes the complex network model at epoch k. In the end of the analysis, a sequence of connection matrices will be estimated, each one associated with an epoch. Studying the network parameters (characteristic path length, clustering coefficient, global efficiency, small worldness) of the single connection matrix, one can analyze the brain connectivity in that specific epoch. The analysis of the entire sequence of connections matrices $\mathbf{CM}^k$ provides a view of the evolution of the complex network parameters over the time, following the dynamics of the EEG recording.

14.2.3   Coherence-Based Complex Network Analysis

A complex network can be mathematically modelled by a graph representation. A graph is a diagram in which objects, called *nodes* or *vertices* (V), are connected to each other through links, called *edges*. Two nodes connected by an edge are *adjacent* or *neighbors* and the number of neighbors represents the *degree* (k) of a node [17]. Each edge can be characterized by a weight, that quantifies the strength of the relationship between the two involved nodes, and a direction, that indicates the path from one node to another, forming a direct graph. The information of a weighted graph is contained in the *nxn connection matrix* (**CM**) where n is the number of vertices and the element **CM***ij* represents the weight of the edge connecting the nodes i and j. In this study, undirected weighted networks were considered. For such networks, the connection matrix is symmetrical. In order to evaluate the efficiency of the model, standard complex metrics were calculated: *Characteristic Path Length* (λ), *Clustering Coefficient* (CC), *Global Efficiency* (GE) and *Small-World property* (SW) [18]. The following definitions refer to weighted networks. The Characteristic Path Length (λ) is a measure of integration of the network [18]. It is the mean of the shortest path length ($d_{i,j}^w$) computed over all the pairs of n nodes and is defined as follows:

$$\lambda = \frac{1}{n(n-1)} \sum_{\substack{i \neq j \\ i,j \in V}} d_{i,j}^w$$

Related to λ is the Global Efficiency (GE) [19]: it is the average inverse shortest path length ($d_{i,j}^w$) and is defined as:

$$GE^w = \frac{1}{n} \sum_{i \in V} \frac{\sum_{j \in V, j \neq i} (d_{i,j}^w)^{-1}}{n-1}$$

The average Clustering Coefficient (CC) is a measure of segregation and quantifies the capability of nodes to group together with the nearest neighbors. It is defined as follows:

$$CC^w = \frac{1}{n} \sum_{i=1}^{n} CC_i^w = \frac{1}{n} \sum_{i \in V} \frac{2t_i^w}{k_i(k_i - 1)}$$

where CC_i represents the clustering coefficient of the node i. It depends on the number of triangles (t_i) around node i and the maximum possible number of triangles of that node that is equal to $k_i * (k_i - 1)/2$, where k_i is the degree of node i [20]. The Clustering coefficient (CC) and the characteristic path length (λ) are the main features of the "small-world" phenomenon [21]. The small-worldness is quantified by the small-world coefficient (SW) which depends on λ, CC (defined previously) and the same parameters (λ_r and CC_r) computed for a random graph of n nodes and E edges. It is defined as:

$$SW = \frac{CC/CC_r}{\lambda/\lambda_r}$$

where $\lambda_r = \frac{ln(n)}{ln(\frac{E}{n}-1)}$ and $CC_r = \frac{(\frac{E}{n})}{n}$.

14.3   Results

In this section, the results achieved applying the method described in Sect. 14.2 will be discussed. The method was applied to the entire set of 10 EEG recordings, including 63 absence seizures. Figure 14.1 shows, as an example, the EEG of channel Fp1 and the behavior of the network parameters estimated for patient pt18. The EEG of this patient shows 6 seizures and two short paroxysms (all marked by vertical lines in Fig. 14.1). The seizures occurred at epochs: 4, 165, 306, 470, 642, 891, whereas, the two short paroxysms occurred at epochs 120 and 448. It is worth to note that, as a result of the abnormal synchronization between EEG signals associated with seizure onset, lambda decreased and the GE, average CC and SW increased. Ponten et al. [11] computed the lambda and the avg CC and observed the same behavior, however, they selected only one seizure per patient. The same behavior occurred in every EEG of the dataset analyzed in the present work. Since the proposed coherence-based complex network analysis was shown to be sensitive to ictal events, the next step of the study was analysing the sequence of connection matrices **CM** preceding the ictal state, in order to look for possible recurring connection patterns related to seizure onset. In fact, epileptic seizures are hypothesized to be triggered by abnormal recruitment phenomena that arise from the focal area (in case of focal seizures) or from all the brain areas simultaneously (in case of generalized seizures). Absence seizures are considered generalized seizures, however, there is evidence that the frontal areas are particularly critical, nonetheless, the recruitment mechanism is still largely unknown. The goal of the present paper was to detect possible patterns of channels recruitment, in order to understand the mechanism of seizure development. Figure 14.2 shows an explanatory sequence of connection matrices for patient pt18 (seizure 3). The coherence values are coded with a coloration ranging from dark grey (coherence = 0) to light grey (coherence = 1). It is worth to point out that electrodes Fp1 and Fp2 showed high coherence values as early as 8 s before the seizure onset and that coherence between O1 and T5 increased 3 s before the onset of the seizure. Soon after seizure onset, a progressive involvement of the other channels took place, starting from T5–T3, T5–F4 and O1–O2 (Fig. 14.2, subplot "seizure"), and then extending the involvement to the entire frontal, temporal and occipital zones (Fig. 14.2, subplot "seizure (Sects. 14.3 and 14.4)").

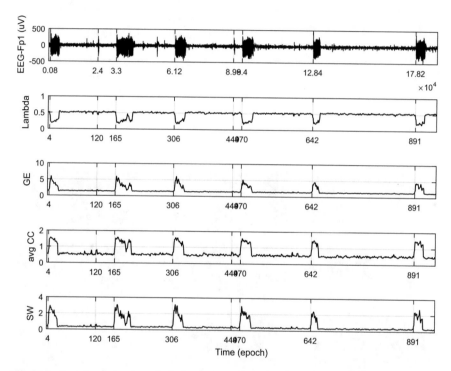

Fig. 14.1 Complex network analysis for patient pt18. The subplot on top displays the EEG of channel Fp1. The remaining subplots show the behavior of the network parameters. The EEG of this patient included 6 seizures and two short paroxysms (all marked by vertical lines in figure). The seizures occurred at epochs: 4, 165, 306, 470, 642, 891, whereas, the two short paroxysms occurred at epochs 120 and 448

Table 14.1 Given a patient, the electrodes that resulted actively connected (Coher ≥ 0.5) in every ictal event are reported in the table

Patient	Critical electrode pairs
13	Fp1 F3, F4 F8, C3 F3
16	T4 T6, T6 C3
18	Fp1 F2, O1 O2, T3 T4, T3 T5
23	T3 T5
29	Fp1 Fp2, T5 T6, O1 O2, Fp1 F7
31	All pairs involving: Fp1, Fp2, F3, F4
32	Fp1 Fp2, F3 Fz
39	Fp1 Fp2, O1 O2, P4 O2, F3 F4, F7 F8
47	Fp1 Fp2 O1 O2, T3 T4
57	Fp1 Fp2 O1 O2, T5 T6

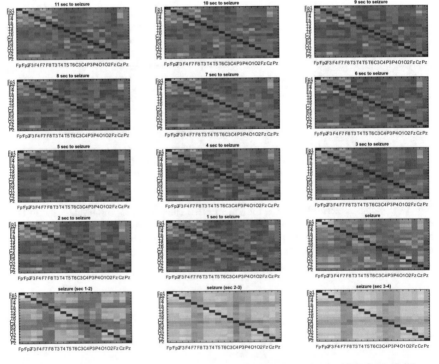

Fig. 14.2 Sequence of coherence-based connection matrices for patient pt18 (seizure 3). Every matrix corresponds to a specific epoch (i.e. EEG window). Every subplot is entitled with the corresponding epoch. The coherence values are coded with a coloration ranging from dark grey (coherence = 0) to light grey (coherence = 1)

Figure 14.3 shows the sequence of connectivity maps corresponding to the connection matrices shown in Fig. 14.2. Such maps display the active links when the connection matrix is thresholded at 0.5 (Coher $\geq$ 0.5). The frontal electrodes look connected as early as 8 s before seizure onset and the occipital electrodes look progressively involved. When the seizure starts, the number of active links increases progressively throughout the ictal event. The same type of inspection was conducted on every seizures of every patient and the results are summarized in Table 14.1 that reports the pairs of electrodes "Critical Electrode Pairs", that resulted actively connected (Coher $\geq$ 0.5) in every ictal event. In conclusion, despite absence seizures are considered "generalized" seizures, the following results were achieved: (1) The involvement of the brain areas appears progressive; (2) The most synchronized channels vary from patient to patient, although the frontal electrodes are prominent; (3) Every subject exhibited peculiar recurrent patterns of area activation.

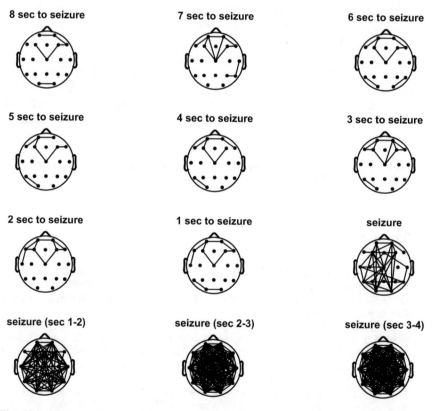

Fig. 14.3 Sequence of connectivity maps corresponding to the connection matrices shown in Fig. 14.2. Every map corresponds to a specific epoch. Every map displays the active links when the connection matrix is thresholded at 0.5 (Coher $\geq$ 0.5)

14.4 Conclusions

In the present paper, the epileptic brain was studied as a complex system by investigating its output: the EEG signals. A dataset of 10 EEG recorded from absence seizure epileptic patients was analyzed. The synchronization strength between the EEG signals was quantified in order to build a complex network model of the dynamical interactions between the cortical areas, whose electrical activity is taken into account by the EEG signals. Synchronization was estimated by means of coherence, due to the promising results it provided on absence seizures EEG analysis in the literature. A coherence-based complex network model was developed in order to assess if the onset of epileptiform activity affected the features of the model. Due to the abnormal synchronization between EEG signals that comes with seizure onset, during the ictal states the Characteristic Path Length decreased whereas the Global Efficiency, the average Clustering Coefficient and the Small Worldness increased. The temporal sequences of the connection matrices preceding the ictal states were also inspected, for every patient, in order to investigate how the channels (i.e. the cortical areas) interact with each other and get involved in the abnormal recruitment mechanism that generates the seizures. The proposed analysis showed that the involvement of the brain areas appears progressive; that the most synchronized channels vary from patient to patient, even though the frontal electrodes are prominent, and that every subject exhibited peculiar recurrent patterns of area activation. Since the achieved preliminary results are encouraging, the analysis will be extended in the near future to a larger dataset.

Acknowledgements Nadia Mammone's work was funded by the Italian Ministry of Health, Project Code GR-2011-02351397.

References

1. Duun-Henriksen, J., Kjaer, T.W., Madsen, R.E., Remvig, L.S., Thomsen, C.E., Sorensen, H.B.: Channel selection for automatic seizure detection. Clin. Neurophysiol. **123**(1), 84–92 (2012)
2. Bandt, C., Pompe, B.: Permutation entropy: a natural complexity measure for time series. Phys. Rev. Lett. **88**(17) (2002)
3. Cao, Y., Tung, W.-W., Gao, J.B., Protopopescu, V.A., Hively, L.M.: Detecting dynamical changes in time series using the permutation entropy. Phys. Rev. E **70**(046217), 1–7 (2004)
4. Bruzzo, A.A., Gesierich, B., Santi, M., Tassinari, C.A., Birbaumer, N., Rubboli, G.: Permutation entropy to detect vigilance changes and preictal states from scalp EEG in epileptic patients. A preliminary study. Neurol Sci. **29**, 3–9 (2008)
5. Ouyang, G., Li, J., Liu, X., Li, X.: Dynamic characteristics of absence EEG recordings with multiscale permutation entropy analysis. Epilepsy Res. **104**(3), 246–252 (2013)
6. Zhu, G., Li, Y., Wen, P.P., Wang, S.: Classifying epileptic EEG signals with delay permutation entropy and multi-scale k-means. Adv. Exp. Med. Biol. **823**, 143–157 (2015)
7. Mammone, N., La Foresta, F., Morabito, F.C.: Discovering network phenomena in the epileptic electroencephalography through permutation entropy mapping. In: Frontiers in Artificial Intelligence and Applications, Neural Nets WIRN10, pp. 260–269 (2011)

8. Mammone, N., Labate, D., Lay-Ekuakille, A., Morabito, F.C.: Analysis of absence seizure generation using EEG spatial-temporal regularity measures. Int. J. Neural Syst. **22**(6) (2012)
9. Mammone, N., Henriksen, J.D., Kjaer, T.W., Morabito, F.C.: Differentiating interictal and ictal states in childhood absence epilepsy through permutation renyi entropy. Entropy **17**(7), 4627–4643 (2015)
10. Mammone, N., La Foresta, G., Inuso, F., Morabito, F.C., Aguglia, U., Cianci. V.: Algorithms and topographic mapping for epileptic seizures recognition and prediction. Front. Artif. Intell. Appl. **204**, 261–270 (2009)
11. Ponten, S.C., Douw, L., Bartolomei, F., Reijneveld, J.C., Stam, C.J.: Indications for network regularization during absence seizures: weighted and unweighted graph theoretical analyses. Exp. Neurol. **217**(1), 197–204 (2009)
12. Rotondi, F., Franceschetti, S., Avanzini, G., Panzica, F.: Altered EEG resting-state effective connectivity in drug-nave childhood absence epilepsy. Clin. Neurophysiol. **127**(2), 1130–1137 (2016)
13. Ravish, D.K., Shenbaga Devi S., Krishnamoorthy, S.G.: Wavelet analysis of EEG for seizure detection: coherence and phase synchrony estimation. Biomed. Res. (2015)
14. Rodrigues, A.C., Machado, B.S., Caboclo, L.O.S.F., Fujita, A., Baccaia, L.A., Sameshima, K.: Source and sink nodes in absence seizures. **2016-October**, 2814–2817 (2016). Cited by 0
15. Gotman, J., Ives, J.R., Gloor, P.: Frequency content of EEG and EMG at seizure onset: possibility of removal of EMG artefact by digital filtering. Electroencephalogr. Clin. Neurophysiol. **52**(6), 626–639 (1981)
16. Welch, P.D.: The use of Fast Fourier Transform for the estimation of power spectra: a method based on time averaging over short, modified periodograms. IEEE Trans. Audio Electroacoust. **15**(2), 70–73 (1967)
17. Chapela, V., Criado, R., Moral, S., Romance, M.: Mathematical foundations: complex networks and graphs (a review). In: Intentional Risk Management through Complex Networks Analysis, pp. 9–36. Springer (2015)
18. Fornito, A., Zalesky, A., Bullmore, E.: Fundamentals of Brain Network Analysis. Academic Press (2016)
19. Latora, V., Marchiori, M.: Efficient behavior of small-world networks. Phys. Rev. Lett. **87**(19), 198701 (2001)
20. Rubinov, M., Sporns, O.: Complex network measures of brain connectivity: uses and interpretations. Neuroimage **52**(3), 1059–1069 (2010)
21. Watts, D.J., Strogatz, S.H.: Collective dynamics of small-world networks. Nature **393**(6684), 440–442 (1998)

Chapter 15
Evolution Characterization of Alzheimer's Disease Using eLORETA's Three-Dimensional Distribution of the Current Density and Small-World Network

Giuseppina Inuso, Fabio La Foresta, Nadia Mammone, Serena Dattola and Francesco Carlo Morabito

Abstract Alzheimer's disease (AD) is the most common neurodegenerative disorder characterized by cognitive and intellectual deficits and behavior disturbance. The electroencephalogram (EEG) has been used as a tool for diagnosing AD for several decades. In the pre-clinical stage of AD, no reliable and valid symptoms are detected to allow a very early diagnosis. There are four different stages associated with AD. The first stage is known as Mild Cognitive Impairment (MCI), and corresponds to a variety of symptoms which do not significantly alter daily life. In the mild stage, an impairment of learning and memory is usually notable. The next stages (Mild and Moderate AD) are characterized by increasing cognitive deficits and decreasing independence, culminating in the patient's complete dependence on caregivers and a complete deterioration of personality (Severe AD). In this paper, we propose the study of the evolution of Alzheimer's disease using eLORETA's three-dimensional distribution of the current density and Small-world network. Our goal is to see the changes of MCI patients' EEG (called EEG T_0) after three months (EEG T_1). The results show that small-world is a valid technique to see the temporal evolution of the disease.

Keywords LORETA · Small-world network · MCI converted to AD

G. Inuso (✉) · F. La Foresta · S. Dattola · F. C. Morabito
DICEAM—Università degli Studi Mediterranea di Reggio Calabria Feo di Vito,
89100 Reggio Calabria, Italy
e-mail: giuseppina.inuso@unirc.it

N. Mammone
IRCCS Centro Neurolesi Bonino-Pulejo, Via Palermo c/da Casazza, SS. 113, Messina, Italy

© Springer International Publishing AG, part of Springer Nature 2019
A. Esposito et al. (eds.), *Quantifying and Processing Biomedical and Behavioral Signals*, Smart Innovation, Systems and Technologies 103,
https://doi.org/10.1007/978-3-319-95095-2_15

15.1 Introduction

A single EEG electrode provides an estimate of synaptic action averaged over tissue masses containing 1 billion neurons. The space averaging of brain potentials resulting from extracranial recording is a data reductions process forced by current spreading in the head volume conductor.

We were interested in monitoring stage at Mild Cognitive Impairment (MCI) to Alzheimer's disease (AD) using the eLORETA's three-dimensional distribution of the current density [1–4] and Small-world network. The technique proposed in this paper aims to describe such changes of an initial EEG (called EEG T_0) after three months (EEG T_1) to study the evolution of the disease. The first EEGs are all MCI's subjects.

The novelty of the paper is represented by the type of evaluation made on the database. We would like to evaluate the evolution of the disease from T_0 to T_1 using markers, which are the core measures of graph theory [5].

Notably, EEG abnormalities of AD patients are characterized by slowed mean frequency, less complex activity, and reduced coherences among cortical regions. These abnormalities suggest that the EEG has utility as a valuable tool for differential and early diagnosis of AD.

Many studies have shown that the graph theory provides an excellent tool to characterize neuronal network capacities for coupling parameters of time-varying signals obtained during resting state EEG [5–7].

15.2 Methodology

EEG functional connectivity analysis was conducted by exact low resolution electromagnetic tomography eLORETA [1] software. This is a linear inverse solution algorithm for EEG with no localization error to point sources under ideal (noise-free) conditions [2].

According to the LORETA method, the current density is the intracortical distribution of the electric activity. Only such a solution can be considered as a tomography in the same sense used in radiological procedures.

The inverse problem is aimed at the reconstruction of sources $\hat{J}$ from electrical signals measured on the scalp, that is the potential $\mathbf{\Phi}$:

$$\hat{\mathbf{J}} = \mathbf{T} * \mathbf{\Phi} \tag{1}$$

where $\mathbf{T}$ (generalized inverse form of $\mathbf{K}$) is a matrix of size [3M $\times$ N] such that

$$\mathbf{K} * \mathbf{T} = \boldsymbol{H}_n \text{ with } \boldsymbol{H}_n = I_n - \frac{1}{N \cdot 1_n \cdot 1_n^T} \tag{2}$$

In Eq. 1, $\boldsymbol{\Phi} \in R^{N_E \times 1}$ is a vector containing scalp electric potentials measured at N_E cephalic electrodes, with respect to a common, arbitrary reference electrode located anywhere on the body.

$\mathbf{K} \in R^{N_E \times 3N_v}$ is the lead field.

Note that the element of this matrix K is $\mathbf{k}_{i,l} = \left(k_{i,l}^x, k_{i,l}^y, k_{i,l}^z\right)$, where $k_{i,l}^x$ is the scalp electric potential at the i-th electrode, due to a unit strength X-oriented dipole at the l-th voxel (i.e. a voxel represents a value on a regular grid in three-dimensional space); $k_{i,l}^y$ is the scalp electric potential at the i-th electrode, due to a unit strength Y-oriented dipole at the l-th voxel; and $k_{i,l}^z$ is the scalp electric potential at the i-th electrode, due to a unit strength Z-oriented dipole at the l-th voxel.

The centering matrix $\mathbf{H}$ in Eq. 2 is the average reference operator.

Equation 1 is the explicit solution to this minimization problem. LORETA is zero error localization.

Output of LORETA will be input of small-world networks.

Connectivity analysis was obtained by Lagged Linear Coherence algorithm as a measure of functional physiological connectivity.

Lagged Linear Coherence in the frequency band ω is defined by the equation

$$\text{Lag} R_{xyw}^2 = \frac{[Im \ Cov(x, y)]^2}{Var(x) * Var(y) - [Re \ Cov(x, y)]^2} \tag{3}$$

where x and y are time series of two Brodmann's areas, Im and Re are imaginary and real part, respectively, Var and Cov are variance and covariance.

Brain connectivity analysis shows considerable promise for understanding how neural pathways gradually break down in aging and Alzheimer's disease.

Core measures of graph theory were computed from http://www.brain-connectivity-toolbox.net [5]. It is a MATLAB toolbox for complex-network analysis of structural and functional brain-connectivity data sets. Segregation refers to the degree to which network elements form separate clusters and corresponds to the clustering coefficient (C) [5]. Integration refers to the capacity of network to become interconnected and exchange informations [8], and it is defined characteristic path length (L) coefficient [5].

The clustering (C) around a vertex i is quantified by the number of triangles in which vertex participates, normalized by the maximum possible number of such triangles [9].

The analysis of brain networks is made feasible by the development of new imaging acquisition methods as well as new tools from graph theory and dynamical systems.

A small-world network is defined to be a network where the typical distance L between two randomly chosen nodes (the number of steps required) grows proportionally to the logarithm of the number of nodes N in the network.

Now we give the definition of the above parameters.

The weighted clustering coefficient C is

$$C = \frac{1}{n} \sum_{i \in N} \frac{2t_i^w}{k_i(k_i - 1)} \quad \text{with } t_i = \frac{1}{2} \sum_{i,j,h \in N} \left(w_{ij} w_{ih} w_{jh}\right)^{1/3} \tag{4}$$

In Eq. 4 ti is the geometric mean of triangles around i; wij are connection weights associated to links (i, j) of graph, assuming that weights are normalized, such that $0 \le wij \le 1$ for all i and j.

The weighted characteristic path length L is defined as

$$L = \frac{1}{n} \sum_{i \in N} \frac{\sum_{j \in N j \neq i} d_{ij}^w}{n - 1} \quad \text{with } d_{ij}^w = \sum_{a_{uv} \in g_{i \leftrightarrow j}^w} f(w_{uv}) \tag{5}$$

In Eq. 5 is represented the shortest weighted path length between i and j; f is a map from weight to length and $g_{i \leftrightarrow j}^w$ is the shortest weighted path between i and j. After the computation of the parameters C and L, we can find the Small-worldness (Sw) parameter.

Sw is defined as the ratio between normalized C and L and Crand and Lrand (parameter of random network with same number of nodes, links and weights) with respect to the frequency bands.

$$S_w = \frac{C/C_{rand}}{L/L_{rand}} \tag{6}$$

The Sw coefficient describes the balance between local connectedness and global integration of a network. If Sw is larger than 1, then a network has small-world properties.

15.3 Results

This section is organized as follows: it describes the database and the result on connectivity measures of small-world network.

15.3.1 Data Description

The analyzed dataset consists of 8 EEG recordings with 19 channels (according to the International 10–20 system), from patients affected by MCI during first recording (T_0). The EEG was high-pass filtered at 0.5 Hz and low-pass filtered at 70 Hz. The sampling rate was set at 256 Hz. The artifacts so have been eliminated using a pre-processing. In fact every type of artifact is typical of a certain frequency range [10, 11].

The EEGs were recorded at IRCCS Centro Neurolesi Bonino-Pulejo of Messina. Patients group included 8 MCI at T_0, and during the control after three months (i.e. T_1) only 3 patients are converted to AD, according to the medical team. Each subject, in fact, was visited by expert neurologists and underwent complete clinical tests [12].

The mean age was about 70 years. There were men and women in this group.

The patients were evaluated according to a protocol approved by the local Ethics Committee and they all signed an informed consent form.

15.3.2   Analysis of Result

For each subject, brain connectivity was computed in 84 regions of interest (ROIs) defined according to the available 42 Brodmann's areas for each of EEG frequency bands (delta (2–4 Hz), theta (4–8 Hz), alpha1 (8–10.5 Hz), alpha2 (10.5–13 Hz), beta1 (13–20 Hz), beta2 (20–30 Hz)).

The hallmark of EEG abnormalities in AD patients is a shift of the power spectrum to lower frequencies (theta and delta band) and a decrease in coherence of fast rhythms (alpha and beta). In particular, many studies have shown that there are nonlinear alterations in the EEG of AD patients, i.e. a decreased complexity of EEG patterns and reduced information transmission among cortical areas, and their clinical implications.

Among other connectivity measures showing disease effects, network nodal degree, normalized characteristic path length, and efficiency decreased with disease, while normalized small-worldness increased, in the whole brain and in left and right hemispheres individually. The normalized clustering coefficient also increased in the whole brain; we discuss factors that may cause this effect.

In particular, the LORETA transform computes the full three-dimensional current density distribution (the current density field). However, the data output of LORETA consists of average current density within predefined locations or ROIs in the source space.

This output will be used as input to study the connectivity measures of small-world network.

In Figs. 15.1 and 15.2, we show the results obtained. The black bar of the histogram is the trend at T_0, and the grey bar is at T_1.

Mostly, horizontal axis of the graph represents specific categories, in our case they are cerebral rhythms (delta, theta, alpha1, alpha2, beta1, beta2), and vertical axis shows the discrete numerical values, i.e. C, L, Sw, K, E, respectively.

C is the parameter of clustering coefficient. It is the fraction of triangles around a node and is equivalent to the fraction of node's neighbours that are neighbours of each other. The characteristic path length L is the average shortest path length in the network. Sw is the coefficient of small-world. K is the density; it is equal to 1. Density is the fraction of present connections to possible connections. E is the parameter of global efficiency of network. The global efficiency is the average inverse shortest path length in the network [5].

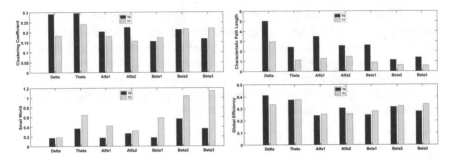

Fig. 15.1 On the top graphics representation there are the clustering coefficient (left) and the characteristic path length (right) for patient 30, that is a MCI both at T0 (black bar) and at T1 (grey bar). On the bottom figures there are the small-world (left) and the global efficiency (right). In horizontal axis there are cerebral rhythms

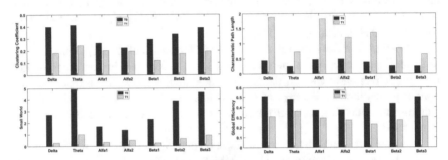

Fig. 15.2 On the top graphics representation there are the clustering coefficient (left) and the characteristic path length (right) for patient 51, that is a MCI at T0 (black bar) as at T1 (grey bar) is an AD. On the bottom figures there are the small-world (left) and the global efficiency (right). In horizontal axis there are cerebral rhythms

They were obtained using the toolbox's functions (weight_conversion; clustering_coef_wu; distance_wei; density_und; efficiency_wei) [5] and their theoretical references have been outlined in the previous section.

We show the graphical representation of patient n.30 and patient n.51 of our database. Patient n.30 is MCI at the time T_0 and at the time T_1, as patient 51 from T_0 to T_1 becomes AD.

The results obtained for all patients in the database reflect the patient's category reported here.

Note that for all patients the networks at all rhythms are highly related because density is equal to 1. These networks have different small-world properties than those of healthy subjects. In fact, the Sw values are almost all below the unit. This means that the two peculiarities of the small-world property, i.e. high clustering coefficient (C) and low path length (L), do not coexist. If a small-world property is missing in a cerebral network then a slower diffusion of information occurs. In fact, the main feature of a small-world network is the presence of shortcuts, which create shortcuts between the nodes, making communication very fast.

In patient 30 (MCI), from time T_0 to time T_1, there is an increase in Sw in all bands, especially at high frequency (beta). In global efficiency, however, there is a reduction in delta and alpha2 bands.

In patient 51 (MCI→AD), from time T_0 to time T_1, there is a decrease of Sw in all bands, especially in theta band. Global efficiency is reduced everywhere, especially in delta.

Based on these results, the reduction of the small-world property and of the global efficiency from T_0 to time T_1 could justify the cognitive decline of the patients; it is due to a reduction in the flow of information between the network elements.

This reduction involves the delta and theta bands, whose variations are mainly associated with memory disorders (i.e. temporal lobe and temporal-parietal lobe). In fact, the hallmark of EEG abnormalities in AD patients is a shift of the power spectrum to lower frequencies (theta and delta band) and a decrease in coherence of fast rhythms (alpha and beta).

15.4 Conclusion

In this paper, we propose the study of the evolution of Alzheimer's disease building a Small-world network using eLORETA's three-dimensional distribution of the current density.

At time T_0 all patients are MCI and at T_1 (3 months after the first EEG recording) only three of the patients convert to AD. We showed the results for MCI patient 30 and for patient 51, who converts to AD.

In MCI patient, from time T0 to time T1, there is an increase in Sw in all bands, especially at high frequency (beta), as in MCI → AD patient there is a decrease of Sw in all bands, especially in theta band, as already known in the literature.

Observing global efficiency there is a reduction in delta band for both patients.

Future research will be devoted to the analysis of the high-resolution EEG using this methodology.

Acknowledgements The authors would like to thank the doctors of IRCCS Centro Neurolesi Bonino-Pulejo of Messina (Italy) for their insightful comments and suggestions.

References

1. Pascual-Marqui, R.D., Lehmann, D., Koukkou, M., Kochi, K., Anderer, P., Saletu, B., Tanaka, H., Hirata, K., John, E.R., Prichep, L., Biscay-Lirio, R., Kinoshita, T.: Assessing interactions in the brain with exact low-resolution electromagnetic tomography. Philos. Trans. A Math. Phys. Eng. Sci. **369**, 3768–3784 (2011)

2. Pascual-Marqui, R.D.: Standardized low-resolution brain electromagnetic tomography (sLORETA): technical details. Methods Find Exp. Clin. Pharmacol. **24**(Suppl D), 5–12 (2002)
3. Cacciola, M., Morabito, F.C., Polimeni, D., Versac, M.: Fuzzy characterization of flawed metallic plates with eddy current tests. Prog. Electromagnet. Res. **72**, 241–252 (2007)
4. Cacciola, M., La Foresta, F., Morabito, F.C., Versaci, M.: Advanced use of soft computing and eddy current test to evaluate mechanical integrity of metallic plates. NDT and E Int. **40**(2), 357–362 (2007)
5. Rubinov, M., Sporns, O.: Complex network measures of brain connectivity: uses and interpretations. Neuroimage **52**, 1059–1069 (2010)
6. Miraglia, F., Vecchio, F., Bramanti, P., Rossini, P.M.: EEG characteristics in "eyes-open" versus "eyes-closed" conditions: small-world network architecture in healthy aging and age-related brain degeneration. Clin. Neurophisiol. **127**, 1261–1268 (2016)
7. Babiloni, C., Frisoni, G.B., Vecchio, F., Pievani, M., Geroldi, C., De Carli, C., Ferri, R., Vernieri, F., Lizio, R., Rossini, P.M.: Global functional coupling of resting EEG rhythms is related to white-matter lesions along the cholinergic tracts in subjects with amnesic mild cognitive impairment. J. Alzheimers Dis. **19**, 859–871 (2010)
8. Sporns, O.: Structure and function of complex brain networks: dialogues. Clin. Neurosci. **15**, 247–262 (2013)
9. Onnela, J.P., Saramaki, J., Kertesz, J., Kaski, K.: Intensity and coherence of motifs in weighted complex networks. Phys. Rev. E Stat. Nonlinear Soft. Matter Phys. **71**, 065103 (2005)
10. Azzerboni, B., Finocchio, G., Ipsale, M., La Foresta, F., Mckeown, M.J., Morabito, F.C.: Spatio-temporal analysis of surface electromyography signals by independent component and time-scale analysis. Proc. Second Joint Meet. IEEE Eng. Med. Biol. Biomed. Eng. Soc. **1**, 112–113 (2002)
11. Mammone, N., Inuso, G., La Foresta, F., Morabito, F.C.: Multiresolution ICA for artifact identification from electroencephalographic recordings. Lect. Notes Artif. Intell. **4692**, 680–687 (2007). (Springer)
12. Labate, D., La Foresta, F., Palamara, I., Morabito, G., Bramanti, A., Zhang, Z., Morabito, F.C.: EEG complexity modifications and altered compressibility in mild cognitive impairment and Alzheimer's disease. Smart Innov. Syst. Technol. **26**, 163–173 (2014). https://doi.org/10.100 7/978-3-319-04129-2_17

Chapter 16
Kendon Model-Based Gesture Recognition Using Hidden Markov Model and Learning Vector Quantization

Domenico De Felice and Francesco Camastra

Abstract The paper presents a dynamic gesture recognizer, that assumes that the gesture can be described by Kendon Gesture model. The gesture recognizer has four modules. The first module performs the feature extaction, using the skeleton representation of the body person provided by NITE library of Kinect. The second module, formed by Learning Vector Quantization, has the task of individuating the initial and the final handposes of the gesture, i.e., when the gesture starts and terminates. The third unit performs the dimensionality reduction. The last module, formed by a discrete Hidden Markov, perfoms the gesture classification. The proposed recognizer compares favourably, in terms of accuracy, most of existing dynamic gesture recognizers.

Keywords Gesture recognition · Kendon model · Hidden Markov Model
Learning Vector Quantization · Principal Component Analysis

16.1 Introduction

Gesture is one of the ways that humans use for exchanging information. For this reason, gesture recognition is a relevant topic in Human Computer Interaction (HCI) discipline. Gestures can be of two different types, i.e., *dynamic gestures* and static gestures, usually called *hand poses* [1]. Hand poses refer to the shape and the orienta-

The research was entirely developed when Domenico De Felice, as M. Sc. student in Applied Computer Science, was at the Department of Science and Technology, University of Naples Parthenope.

D. De Felice
University of Brescia, via Valotti 9, 25123 Brescia, Italy
e-mail: domenico.defelice@alice.it; domenico.defelice@unibs.it

F. Camastra (✉)
Department of Science and Technology,
University of Naples Parthenope, Centro Direzionale Isola C4, 80143 Naples, Italy
e-mail: camastra@ieee.org

tion of the hand, whereas dynamic gestures consider the movement and the position of the hand.

Several theoretical models have been proposed for describing human dynamic gestures. Among them, Kendon gesture model is one of the most relevant. This model assumes that each human gesture can be viewed as a sequence of relevant handposes. The aim of the work is the construction of a dynamic gesture recognizer, assuming that the gesture can be described by Kendon model. The recognizer should perform two different tasks. Firstly, the recognizer should individuate when the gesture starts and terminates. Then, the recognizer performs the classification of the so-individuated gesture.

The paper is organized as follows. Section 16.2 introduces the Kendon model of human gesture; Sect. 16.3 describes the Gesture Recognizer; Sect. 16.4 presents some experimental results; finally, in Sect. 16.5 some conclusions are drawn.

16.2 Kendon Model of Gesture

Among models proposed for describing human dynamic gestures, the model proposed by Kendon [2], referred, in the work, as *Kendon Gesture Model*, is the most popular. In the Kendon Gesture Model, any dynamic gesture can be viewed as a sequence of relevant hand poses, called *strokes*. According to this model, a dynamic gesture can be divided in three phases, i.e., *Preparation*, *Nucleus*, and *Recovery*.

In the phase of Preparation, the person body moves from the initial (or rest) position to the one where the gesture will be performed. The second phase, the Nucleus, can be composed of two further steps, namely *Stroke* and *Hold*. In Kendon's point of view the Stroke corresponds to what is commonly identified by people as gesture. Hold denotes the presence of a short pause during the gesture performing. In the last phase, the person comes back to the rest position.

Although there have been some attempts [3], segmenting a dynamic gesture in strokes is a process subject to segmentation errors that can compromise the overall recognition of the gesture. Therefore, in the rest of the paper the Kendon model is used in the simplest way, i.e., it assumes that each dynamic gestures starts with the rest position and terminates with the recovery.

16.3 Gesture Recognizer

The gesture recognizer is composed of four units, the *Feature Extractor*, the *Rest Position and Retraction Identifier*, the *Dimensionality Reduction Module*, and the *Dynamic Gesture Classifier*.

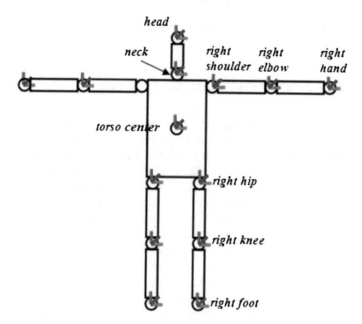

Fig. 16.1 Skeleton representation of the human body provided by NITE library. The names of the joint of the left part of the skeleton are identical to the names of the respective joints of the right part

16.3.1   Feature Extractor

The flux of image frames is acquired by Kinect [4] with a refresh frequency of 50 Hz. Each frame is undergone to the following feature extraction process. The feature extraction process has two stages. The former stage consists in computing, by NITE library [5], the skeleton representation of the person (see Fig. 16.1) whose gesture has to be recognized. The skeleton representation is formed by a set of 16 *joints*. For each joint NITE library provides the position and, sometimes, the orientation, expressed by *quaternions* [6].[1] Then, it selects only the joints involved in dynamic gesture performing, i.e., the ones of arms. Therefore, the following 20 features are extracted:

- 1st-2nd-3th correspond to $JOINT\_LEFT\_HAND$ coordinates
- 4th-5th-6th correspond to $JOINT\_LEFT\_ELBOW$ coordinates
- 7th-8th-9th-10th correspond to $JOINT\_LEFT\_ELBOW$ quaternions
- 11th-12th-13th correspond to $JOINT\_RIGHT\_HAND$ coordinates
- 14th-15th-16th correspond to $JOINT\_RIGHT\_ELBOW$ coordinates
- 17th-18th-19th-20th correspond to $JOINT\_RIGHT\_ELBOW$ quaternions

---

[1]Recent trackers, prefers to measure orientation by quaternions, instead of usual Euler angles, since Euler angle representation can be affected by the *gymbel lock*.

At the end of the process all the features are normalized.

16.3.2 Rest Position and Retraction Identifier

As pointed out in Sect. 16.2, Kendon Gesture model assumes that any dynamic gesture begins with the rest position and terminates with retraction. The task of this module is the identification of the rest position in the dynamic gestures. The module consists in *Learning Vector Quantization (LVQ)* [7, 8] classifier. LVQ has been chosen since it requires moderate computational resources and it is effective in disparate domains such as the classification of nonstationary power signals [9], and arrythmias [10], and the handpose recognition [11]. LVQ is a supervised vector quantization algorithm that aims to represent the input data by a smaller number of *codevectors* minimizing the misclassification. LVQ consists of the application of two consecutive learning algorithms, LVQ1 and LVQ2 or alternatively LVQ3. LVQ1 training adopts the winner-takes-all rule, comparing the input vector class of with the one of the closest codevector. If the classes are the same, the codevector is approached to the input vector, otherwise it is pushed away. No modifications are applied to the remaining codevectors. LVQ2 improves LVQ1 training by pairwise arrangement of codebook vectors belonging to adjacent classes, yielding a decision surface that ap-

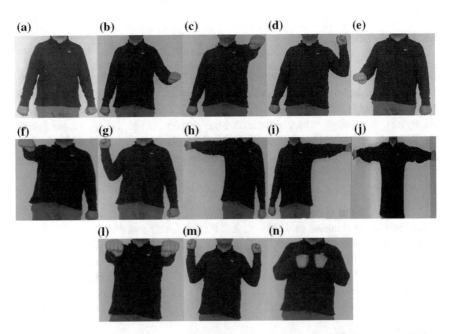

Fig. 16.2 Strokes represented in the database. The person in the figure is the first author of paper, who gives his consent to show his images in the paper

proximates roughly the Bayes' rule decision surface. The reader is remanded to [7, 8] for a more extensive LVQ description. For LVQ training it has been constructed a database formed by 13 different strokes (see Fig. 16.2), manually selected by gestures. Each training was perfomed adopting consecutive LVQ1+LVQ2 as learning sequence and using *LVQ-pak* [7] software library. Several configurations of LVQ net have been trained, changing LVQ1 and LVQ2 parameters, and the best LVQ has been picked by *cross-validation* [12].

Having said that, the LVQ classifier receives for each frame its 20 features representation and individuates the frames where the rest position and retraction occurr. All the frames, included between the rest position and the retraction frames are transmitted to the following modules for performing the dynamic gesture recognition.

16.3.3   Dimensionality Reduction Module

The module receives the sequence of 20 feature vectors, extracted from each frame between the rest position and retraction, and performs a dimensionality reduction process for each feature vector. In order to perform this process, it is used the simplest and the most popular Dimensionality Reduction algorithm, the *Principal Component Analysis* (PCA) [13]. We recall the PCA projects the data along the directions of their maximal variance. The reader is remanded to [13] for a more extensive PCA description.

An analysis of the eigenspectrum computed on feature data has shown that in the first two and five principal components is concentrated about 60 and 90% of the information, respectively. For sake of complexity reduction, it has been decided to consider only the first two principal components. Therefore, each 20 feature vector extracted from the frame is projected onto two first principal components.

16.3.4   Dynamic Gesture Classifier

The Dynamic Gesture Classifier is formed by a *Hidden Markov Model* (HMM) [14, 15]. The HMM classifier has been chosen, since it is used with success in speech and handwriting recognition and it is the most suitable algorithm for classifying sequences of observations of different length. The HMM of the dynamic gesture classifier has discrete output and has a topology left-to-right of Bakis type [14]. A comprehensive survey on Hidden Markov Models can be found in [14].

For the construction of the discrete HMM has been used the framework *Xkin* [16], properly modified. The training of the model has been performed by *Baum-Welch algorithm* [14, 17]. Several configurations of the discrete HMM have been tested, changing model parameters, e.g., the number of hidden states. The best discrete HMM has been selected by *crossvalidation* [12].

Table 16.1 The confusion matrix of dynamic gesture recognizer, without rejection, on the test set. The values are expressed in terms of percentage rates

	#1	#2	#3	#4	#5	#6	#7	#8
#1	96	4	0	0	0	0	0	0
#2	1	99	0	0	0	0	0	0
#3	0	0	94	6	0	0	0	0
#4	0	0	1	99	0	0	0	0
#5	0	0	0	0	99	1	0	0
#6	0	0	0	0	2	98	0	0
# 7	0	0	1	0	0	0	99	0
# 8	0	0	0	0	0	1	0	99

Table 16.2 Gesture recognizer comparison

Recognizer	#Gestures	Accuracy (%)
Sultana and Rajapuspha [19]	5	80.00
Tang [18]	2	96.10
Zhu and Yuan [20]	8	91.12
Our system	8	97.87

16.4 Experimental Results

For the testing of the dynamic gesture recognizer has been collected a database of 800 dynamic gestures, formed by 8 different gestures (see Fig. 16.3), each repeated 100 times. The first two gestures are the same used in Madeo et al's work [3]. The database has been splitted in two equal parts, training and test set. The accuracy of dynamic gesture recognizer on the test set has been 97.87%. The confusion matrix, on the test set, is reported in Table 16.1.

The hand pose recognizer is implemented in C++ under 64 bit Windows 10 Microsoft on a PC with i5–2410 M Processor 2.30 Ghz, 6 GB RAM and requires 462 CPU ms to recognize a single dynamic gesture.

16.4.1 Comparison with Other Systems

The dynamic gesture recognizer proposed has been compared with other existing gesture recognizers of the literature [18–20]. All recognizers considered use the Kinect sensor as acquisition device. The comparison among different recognizers is summarized in Table 16.2.

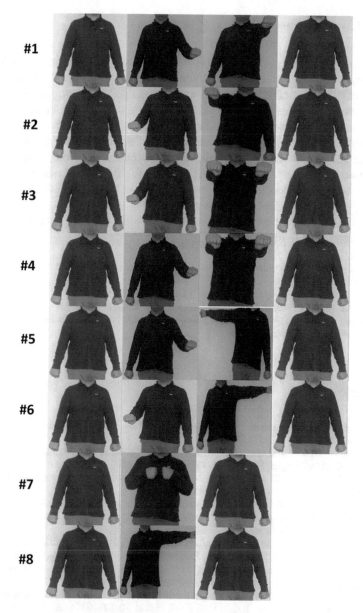

Fig. 16.3 Dynamic gestures in the database. The person in the figure is the first author of paper, who gives his consent to show his images in the paper

16.5   Conclusions

In the paper a dynamic gesture, based on Kendon Model has been presented. Therefore, it has been assumed that each dynamic gesture starts with a rest position and
terminates with a retraction handpose. The proposed dynamic gesture has four stages.
The first stage performs the feature extraction, using the skeleton representation of
the person body provided by NITE library. The second stage individuates the initial and final handposes of gesture, namely when the gesture starts and terminates.
The third one applies the dimensionality reduction process, by computing PCA, on
each feature vector. The last module performs the classification by a discrete Hidden
Markov Model. The proposed recognizer outperforms, in terms of accuracy, most of
dynamic gesture recognizers of the literature.
In the next future we plan to investigate the application of dynamic gesture recognizer
in ambient assisted living domain.

Acknowledgements FIrstly, the authors wish to thank the anonymous reviewers for their valuable
comments.
Part of the work, was developed by Domenico De Felice during his M. Sc. thesis in Computer
Science at University of Naples Parthenope, with the supervision of Francesco Camastra.
This research was funded by *Sostegno alla ricerca individuale per il triennio 2015–17* project of
University of Naples Parthenope.

References

1. Mitra, S., Acharya, T.: Gesture recognition: a survey. IEEE Trans. Syst. Man and Cyber. Part
   C: Appl. Rev. **37**(3), 311–324 (2007)
2. Kendon, A.: How gestures can become like words. In: Crosscultural perspectives in nonverbal
   communication, Toronto, Hogrefe, pp. 131–141 (1988)
3. Madeo, R., Peres, C., de Moraes Lima, S.: Gesture phase segmentation using support vector
   machines. Expert Syst. Appl. **56**, 100–115 (2016)
4. Zhang, Z.: Microsoft kinect sensor and its effect. IEEE Trans. Multimedia **19**(2), 4–10 (2012)
5. Davison, A.: Kinect Open Source Programmimg Secrets: Hacking the Kinect with openNI.
   NITE and Java, Mc Graw Hill Professional (2012)
6. Zhang, F.: Quaternions and matrices of quaternions. Linear Algebra its Appl. **251**, 21–57 (1997)
7. Kohonen, T., Hynninen, J., Kangas, J., Laaksonen, J., Torkkola, K.: Lvq-pak: The learning
   vector quantization program package. In: Technical Report A30, Helsinki University of Technology, Laboratory of Computer and Information Science (1996)
8. Lamberti, L., Camastra, F.: Handy: a real-time three color glove-based gesture recognizer with
   learning vector quantization. Expert Syst. Appl. **39**(12), 10489–10494 (2012)
9. Biswal, B., Biswal, M., Hasan, S., Dash, P.: Nonstationary power signal time series data classification using lvq classifier. Appl. Soft Computing **18**, 158–166 (2014)
10. Melin, P., Amezcua, J., Valdez, F., Castillo, O.: A new neural network model based on the lvq
    algorithm for multi-class classification of arrhythmias. Inf. Sci. **279**, 483–497 (2014)
11. Lamberti, L., Camastra, F.: Real-time hand gesture recognition using a color glove. In: Pro-
    ceedings of the 16th international conference on Image analysis and processing, ICIAP 2011,
    Springer Verlag, pp. 365–373 (2011)
12. Hastie, T., Tibshirani, R., Friedman, R.: The elements of statistical learning. 2nd edn. Springer
    (2009)

13. Jollife, I.T.: Principal component analysis. Springer-Verlag, Berlin (1986)
14. Rabiner, L.: A tutorial on hidden markov models and selected applications in speech recognition. Proc. IEEE **77**(2), 257–286 (1989)
15. Camastra, F., Vinciarelli, A.: Markovian models for sequential data. In: Machine Learning for Audio, Image and Video Analysis, Springer, pp. 295–340 (2015)
16. Pedersoli, F., Adami, N., Benini, S., Leonardi, R.: Xkin: extendable hand pose and gesture recognition library for kinect. In: Proceedings of the 20th ACM International Conference on Multimedia, pp. 1465–1468 (2012)
17. Baum, L., Petrie, T., Soules, G., Weiss, N.: A maximization technique occurring in the statistical analysis of probabilistic functions of markov chains. Ann. Mathe. Statist. **41**, 164–171 (1970)
18. Tang, M.: Recognizing hand gestures with microsofts kinect. In: Technical report, University of Stanford, Department of Electrical Engineering (2011)
19. Sultana, A., Rajapuspha, T.: Vision based gesture recognition for alphabetical hand gestures using the svm classifier. Int. J. Comput. Sci. Eng. Technol. **7**(3), 218–223 (2012)
20. Zhu, Y., Yuan, B.: Real-time hand gesture recognition with kinect for playing racing video games. In: Proceedings of 2014 International Joint Conference on Neural Networks, IEEE, 3240–3246 (2014)

Chapter 17
Blood Vessel Segmentation in Retinal Fundus Images Using Hypercube NeuroEvolution of Augmenting Topologies (HyperNEAT)

Francesco Calimeri, Aldo Marzullo, Claudio Stamile and Giorgio Terracina

Abstract Image recognition applications has been capturing interest of researchers for many years, as they found countless real-life applications. A significant role in the development of such systems has recently been played by evolutionary algorithms. Among those, HyperNEAT shows interesting results when dealing with potentially high-dimensional input space: the capability to encode and exploit spatial relationships of the problem domain makes the algorithm effective in image processing tasks. In this work, we aim at investigating the effectiveness of HyperNEAT on a particular image processing task: the automatic segmentation of blood vessels in retinal fundus digital images. Indeed, the proposed approach consists of one of the first applications of HyperNEAT to image processing tasks to date. We experimentally tested the method over the *DRIVE* and *STARE* datasets, and the proposed method showed promising results on the study case; interestingly, our approach highlights HyperNEAT capabilities of evolving towards small architectures, yet suitable for non-trivial biomedical image segmentation tasks.

Keywords Retinal fundus images · Blood vessel segmentation · Hyperneat
Evolutionary neural networks · Genetic algorithms

The work is partially funded by an European Union's Horizon 2020 research and innovation programme under the Marie Skłodowska-Curie grant agreement No. 690974 and Dottorato innovativo
a caratterizzazione industriale PON R&I FSE-FESR 2014–2020.

F. Calimeri · A. Marzullo (✉) · G. Terracina
Department of Mathematics and Computer Science, University of Calabria, Rende, Italy
e-mail: marzullo@mat.unical.it

F. Calimeri
e-mail: calimeri@mat.unical.it

G. Terracina
e-mail: terracina@mat.unical.it

C. Stamile
Department of Electrical Engineering (ESAT), STADIUS,
Katholieke Universiteit Leuven, Leuven, Belgium
e-mail: Claudio.Stamile@esat.kuleuven.be

© Springer International Publishing AG, part of Springer Nature 2019
A. Esposito et al. (eds.), *Quantifying and Processing Biomedical and Behavioral
Signals*, Smart Innovation, Systems and Technologies 103,
https://doi.org/10.1007/978-3-319-95095-2_17

173

17.1 Introduction

Long-held promise have been made, relying on artificial neural networks, in a number of object recognition benchmarks; among them, for instance, we count the MNIST hand-written digit dataset [1, 2] or the ImageNet Large-Scale Visual Recognition Challenge [3]; quite often, the current best performances on these tasks have been achieved. Besides such impressive results obtained by hand-designed structures, the effectiveness of a relatively novel approach in the area of machine learning has recently started to be investigated in visual processing tasks: Neuroevolution (NE). NE is a form of machine learning that uses evolutionary algorithms to design parameters, topology and structure of Artificial Neural Networks. This kind of approach has been already shown to be successful in many tasks; most notably, it has been effectively applied to benchmarks where the exact target outputs or policy are unknown, and only a scalar evaluation signal is available [4]. Hypercube-based Neuro-Evolution of Augmenting Topologies (HyperNEAT) is a neuroevolution method that uses an indirect encoding – hypercube weight pattern encoding – that makes the generation of large-scale neural networks much easier for the evolutionary algorithm, by taking advantage from the geometric regularities in a potentially high-dimensional input space.

In this work we investigate the effectiveness of HyperNEAT in the automatic pixel-wise classification of blood vessels in retinal fundus images. Interestingly, the application of evolutionary neural networks to image processing has not been widely explored, so far; in particular, the herein proposed approach consists of one of the first applications of HyperNEAT to image processing tasks to date. Moreover, to the best of our knowledge, only few applications of this evolutionary approach on biomedical image processing task were performed so far. A similar technique, namely Feature-Deselective Neuroevolution Classifier (FD-NEAT) [5], has been show to be capable to retain relevant features derived from CT images. Here, the high input size problem was treated, by allowing input nodes to be dropped to reduce the complexity of the input and, consequently, the complexity of evolved networks.

We discuss the results of a proper experimental activity on the *DRIVE*[1] and *STARE*[2] databases, that suggest as HyperNEAT can achieve good level of accuracy in a non-trivial task image processing task.

The remainder of the paper is structured as follows. In Sect. 17.3 we provide a detailed description of our approach, introducing also related literature. In Sect. 17.4 we present our experimental activities, and discuss the results in Sect. 17.5. Finally, in Sect. 17.6 we draw our conclusion.

---

[1]http://www.isi.uu.nl/Research/Databases/DRIVE/.

[2]http://cecas.clemson.edu/~ahoover/stare/.

17.2   Related Work

The HyperNEAT algorithm has been investigated on different image recognition benchmarks [4, 6, 7], where it showed good and encouraging results over other mentioned methods. However, most of these benchmarks consisted of easy pattern recognition tasks, such as identifying the position of single objects in an image, finding the biggest object, or recognizing the direction of curvilinear objects. In [8], however, the usage of HyperNEAT as a feature extractor in the image recognition domain of numbers has been investigated, in the context of Deep Learning architectures. Experiments showed that HyperNEAT itself struggles to find out ANNs that perform well while dealing with image classification tasks; nevertheless, the work in [8] proved that HyperNEAT enjoys the effective ability to act as a feature extractor, and it has been combined with a classical ANN trained by backpropagation.

The same authors of [8] conducted an interesting study where HyperNEAT has been applied to classify maritime vessels from satellite imagery [9]. Results show that HyperNEAT is able to learn features from such imagery that allow improved classification performance, with respect to Principal Component Analysis (PCA); furthermore, HyperNEAT enables the unique capability to scale image sizes through the indirect encoding.

Furthermore, in [10] the authors combined NeuroEvolution method HyperNEAT with Novelty Search for image recognition, suggesting promising results.

The aim of the present work is to explore the capability of HyperNEAT in solving a concrete biomedical problem, namely the segmentation of blood vessels in retinal fundus images. It is worth noting that this task slightly differs from all benchmarks used in previous studies, because of the large size of the input. Furthermore, variation in brightness and illumination of retinal fundus images is a non-trivial issue to be managed by the algorithm.

17.3   Proposed Approach

In the following, we introduce some background techniques and methods, and provide further details on the proposed approach.

17.3.1   NeuroEvolution of Augmenting Topologies (NEAT)

The architecture of a neural network plays a crucial role in making its applications effective. The NEAT algorithm optimizes both structure and weights of a neural network, and attempts to find a solution to several technical challenges such as: (i) provide a genetic representation that allows a meaningful application of the crossover operation; (ii) protect topological innovations by avoiding the "premature" disappearance of a new network from the gene pool, so that it can be optimized; (iii) minimize topologies throughout evolution without the need for a specially contrived fitness function that measures complexity. The NEAT genome structure contains a list of genes representing neurons, *neuron genes*, and connections, *link genes*. A link gene holds information about the two neurons it connects, the weight related to the connection, a flag that indicates if the link is enabled, and other useful informations. A neuron gene holds information about the type of the neuron (input, output or hidden) and the activation function. NEAT makes use of four mutation operations, that include: adding new connections, perturbing a connection weight, adding new hidden nodes, disabling or enabling genes in the chromosome.

17.3.2 Hypercube-Based Neuro-Evolution of Augmenting Topologies (HyperNEAT)

Recently discovered neuroevolution methods have typically only been able to produce networks with small numbers of inputs and outputs. However, we know that many practical applications (image processing, for instance) require methods for evolving neural networks that can directly process high-dimensional input or output spaces, and might contain hundreds or thousands of inputs/outputs [4].

A method which attempts to overcome this limitation has been presented in [11]: HyberNEAT is the indirect encoding extension of NEAT. Therein, instead of applying the evolved network directly to the task, the idea is to evolve a genotype that defines the connections for a possibly much larger substrate network, i.e., the phenotype that is actually applied to the task. This indirect encoding is performed via a Compositional Pattern Producing Network (CPPN) designed to represent patterns of regularity such as symmetry, repetition, and repetition with variation. The CPPN is evolved using an extension to the NEAT algorithm that permits chromosomes to represent the expanded set of node activation functions [4].

The origin of the name HyperNEAT comes from the idea that a CPPN paints a pattern on the inside surface of a hypercube, i.e. a 4-dimensional cube. When represented as a pattern of connectivity in 2-dimensional space, that pattern effectively forms a neural network onto itself [12]. This emergent network is called a *substrate*. Once the topology of the substrate is defined, the CPPN is queried for the value of each weighted connection in the substrate. The CPPN is provided with the coordinates of each node as input, and the output of the CPPN is the weight value for the connection between those nodes (Fig. 17.1).

The main implication of HyperNEAT is that it can efficiently evolve very large neural networks which actually "see" the geometry of the problem domain and can make use of such geometry in order to significantly enhance learning [11].

Fig. 17.1 The CPPN is provided with the coordinates of each node as input, and the output of the CPPN is the weight value for the connection between those nodes

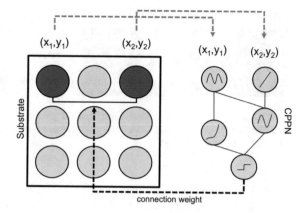

17.3.3   Problem Definition

Segmentation of blood vessels in retinal fundus images plays an important role when dealing with several pathologies, like hypertension, diabetes, and cardiovascular diseases. Accurate vasculature segmentation is, however, a difficult task for several reasons: the presence of noise, the low contrast between vessels and background, the variability of vessel width, brightness, and shape. Many methods for automatic retinal vessel segmentation have been introduced, such as techniques based on vessel tracking, mathematical morphology, matched filtering, model-based locally adaptive thresholding, deformable models or supervised method based on pixel classification. In the last few years, emerging approaches are based on Deep neural networks [2] which achieved impressive results [3, 13].

17.3.4   Model Definition

In this paper, we analyze the effectiveness of HyperNEAT when applied to the task of segmenting blood vessels in digital retinal fundus images. It is worth noting that the application of HyperNEAT in this context is not obvious, due to the huge size of the input; in order to effectively face this issue, a method based on *sliding window* is applied. In more detail, each image is cropped in a grid of sub-images of size $N \times N$, whose dimension depends on the input of the network. Elements of the corresponding set are iteratively fed into the network, and the final image is reconstructed starting from all the partial results, as illustrated in Fig. 17.2. For each sub-image, the label is represented by the intensity value of the central pixel of the corresponding target sub-image. A square HyperNEAT substrate is then defined according to the dimensions of the pre-defined training set: an $N \times N$ grid of neurons as input layer along with a single neuron as output layer.

(a)   (b)   (c)   (d)

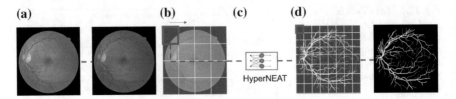

Fig. 17.2 Architecture of the proposed approach: the green channel is extracted (**a**). Then, for each $N \times N$ subimage (**b**), the pixel classification is performed using the evolved network (**c**). Finally, the image is reconstructed (**d**)

Furthermore, we do not apply the HyperNEAT algorithm from scratch: we initialize the genes starting from a network evolved on the task of reproducing binary images; interestingly, we found that, when the initialization relies on this kind of auto-encoder, the population better reacts to high- or low-intensity values. The evolution is first performed on a training set created by extracting random sub images from the original dataset; the best-performing chromosomes are then *fine tuned* on a training set created by extracting sub-images from particular regions of the images (i.e., interesting areas such as optic disc or areas with high illuminance variance). These steps are iterated over multiple runs, in order to produce the final chromosome.

17.4 Experimental Analysis

We discuss next a proper experimental evaluation of the proposed approach.

17.4.1 Dataset Description

The evolution was performed and tested on two public available databases: the Digital Retinal Images for Vessel Extraction (*DRIVE*) database, and The STructured Analysis of the Retina (*STARE*) database. As reported by the official website, the DRIVE database was obtained from 400 diabetic subjects of age ranging in the 25–90 years interval. 40 photographs have been randomly selected: 33 did not show any sign of diabetic retinopathy and 7 showed signs of mild early diabetic retinopathy.

The images were acquired using a Canon CR5 non-mydriatic 3CCD camera with a 45 degree field of view (FOV) and a diameter of 540 pixels, using 8 bits per color plane at 768×584 pixels. For this database, images have been cropped around the FOV, and a mask image is provided that delineates the FOV for each image.

The *STARE* dataset consists of 20 PPM images, digitized slides captured by a TopCon TRV-50 fundus camera with 35 degree field of view. Each slide was digitized to produce a 605×700 pixel image at 24 bits per pixel.

Twenty images from the *DRIVE* database were used to evolve the Neural Network, while forty images from *DRIVE* and *STARE* were used to test the model.

17.4.2 Preprocessing

In order to improve results, we adopted a preprocessing phase. In particular, the green channel is extracted for each image; as a matter of fact, according to consistent findings in many related works, this channel provides the best local contrast between background and foreground [14].

17.4.3 Evolution of the Model

An accurate selection of the best parameters for the evolution of the Neural Network (NN) using the HyperNEAT algorithm is a crucial point for both obtaining the best performances and properly understanding the behavior of the neural network according to each parameter; the evolution of a model that is able to solve the task with a reasonable level of accuracy, indeed, is strictly related to the choice of parameters. We report below some of the crucial parameters which allowed us to significantly improve performances, and how we selected them.

In order to find the best combination of parameters, a grid search has been performed. In more detail, parameters were changed on a logarithmic scale according with the bounds recommended by the authors of the framework used in the experiments [4]. The choice of such ranges is due to the fact that finer step size did not significantly improve the results any further. In our settings we used a population size of 250 and mutation rates to add connections, add neurons, remove connections, remove neurons and weight mutation rates of 0.1, 0.3, 0.3, 0.5, 0.7 respectively. Allowed activation functions for the CPPN were sigmoid, sign, xor, ramp, reciprocal, step, multiply, tanh-cubic and rectified linear unit.

To evaluate the performance of the chromosome during the evolution, two main similarity measures were considered; these have been used as fitness function for the genetic algorithm. The Sørensen-Dice Score Coefficient (DSC) was computed according to: $DSC = \frac{2*|A \cap B|}{|A|+|B|}$, where A is the voxel set containing positive pixels and B is the voxel set with the set of positive pixel classified by our method. One of the main issue we had to take into account is the high level of unbalance between the weak and the strong classes, due to the eminent presence of *non vessel* pixels in the image against the small amount of *vessel* pixels. We dealt with this issue by defining a new performance measure obtained by the normalization of the sum of the Dice coefficient computed both for positive and the negative class.

As already stated above, only the DRIVE dataset has been used to perform the evolution. The grid search was then performed using part of the DRIVE, and the performances were computed on STARE and DRIVE datasets already described in Sect. 17.4.1. It is important to point out that the algorithm never used information derived from the test set during the training phase.

17.4.4 Evaluation

The choice of the performance measure to use by the genetic algorithm depended on a proper analysis of the consequences that any error in the segmentation might have. Some of the best solutions we found consist of *overlap ratio measures*, which appear to work well in many situations: unlike *volume error*, they are sensitive to failure of the segmentation label; however, they are someway unsusceptible to volumetric under- and over-estimations. Shape infidelity is only captured if the deviation is volumetrically impactful: a thin panhandle will not result in a large deviation from one. In order to evaluate results, we used Accuracy ($Acc = \frac{TP+TN}{P+N}$) according to standard methodologies already adopted in related works, where $P + N$ is the total number of pixels in the image.

17.5 Discussion

Figure 17.3 illustrates the quality of the results achieved by our method: it shows the blood vessel segmentation in three images of the retinal fundus from DRIVE and STARE databases. The evolution was performed for 10 runs of 500 generations each, using a training set of 35,000 of 6×6 sub images. Table 17.1 shows the average accuracy of our approach. As one can observe, performances on the two dataset are comparable (about 80% of accuracy on average), with slightly better results for STARE. Observe that, noise in the images guides the system to prefer only strongly evident vessels; as a consequence, it is easier for the system to find true negatives rather than true positives. Given the nature of the images and of the accuracy measure, since the number of true negatives is generally much higher than the number of true positives, system behaviour results in a higher accuracy.

Table 17.2 shows the achieved results with respect to the state-of-the-art methods. The proposed approach does not directly improve the state-of-the-art, actually, however it shows many interesting features which are next explained. As we can observe from graphical results, the shared shape of the blood vessel is precise and well defined. It is worth remarking that the classification of the whole image is, actually, the result of a *sliding-window*-like procedure: this implies that the overall classification is the result of the classification of several sub-images; interestingly, it is worth noting how coherence is preserved at borders among all sub-images. Results show that HyperNEAT can evolve a relatively small suitable structure: we defined

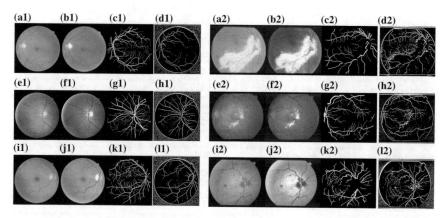

Fig. 17.3 Examples of application of our segmentation method on three images from DRIVE (1) and STARE (2) with different illumination conditions. **a, e, i** Original RGB retinal images. **b, f, l** Preprocessed image. **c, g, m** The manual segmentation results. **d, h, n** Segmentation results obtained by means of the herein proposed method

Table 17.1 Performances in terms of average accuracy (Mean) (± standard deviation in parenthesis), min accuracy (Min) and max accuracy (Max)

	Mean	Min	Max
DRIVE	0.8031 (±0.03)	0.7427	0.9292
STARE	0.8314 (±0.03)	0.7883	0.9205

a substrate of 6×6 input neurons, and just 1 output neuron. Moreover, the CPPN evolved by NEAT contains just 26 neuron genes and 48 connection genes. Convergence is also interesting; Fig. 17.4 depicts the trend of the fitness over the number of generations in both the training and the test phase.[3] As can be easily observed, fitness increases rapidly, reaching a good threshold already in the first steps of the evolutionary process. Furthermore, an interesting coherence between the training and the test phase is observable: when the overall fitness increases, the same perturbation occurs in the test phase, suggesting that HyperNEAT also easily generalizes the classification of previously unseen images. All these considerations suggest that, even if our approach does not directly improve accuracy w.r.t. state-of-the-art, it requires very low computational power yet achieving high levels of accuracy; as a consequence, it might be envisaged its use on portable and mobile devices for fast and accurate diagnoses.

---

[3]Here standard deviation is depicted as a band around the line; observe that an overlapping region is evidenced between the two bands.

Table 17.2 Comparing the segmentation results of different algorithms with our method on DRIVE and STARE database in terms of average Accuracy (ACC)

Method type	Method	DRIVE	STARE
Supervised	Staal et al.	0.9441	–
Supervised	Niemeijer et al.	0.9417	–
Supervised	Soares et al.	0.9466	0.9480
Supervised	Ricci and Perfetti et al.	0.9595	0.9646
Supervised	Marin et al.	0.9452	0.9526
Supervised	**This work**	0.8031	0.8313
Matched filter	Chaudhuri et al.	0.8773	–
Matched filter	Cinsdikici and Aydin	0.9293	–
Ruled-based	Hoover et al.	–	0.9275
Ruled-based	Mendonça et al.	0.9463	0.9479
Ruled-based	Martinez-Perez et al.	0.9344	0.9410
Model-based	Jiang and Mojon	0.8911	0.9009
Clustering	Zhang et al.	0.9400	–

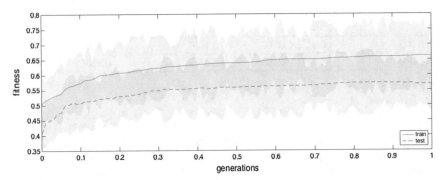

Fig. 17.4 Average best fitness over 30 runs in terms of fitness and number of generations. Shaded areas depict standard deviation

17.6   Conclusion

In this work we presented a method for the automatic segmentation of blood vessels in retinal fundus images, motivated by the known fact that retinal changes reflect systemic microvascular damages associated with a number of pathological conditions, such as hypertension or diabetes. The proposed method is based on HyperNEAT, a genetic algorithm defined to evolve the structure of an artificial neural network. We carried out proper experimental activities aimed at assessing effectiveness and performance of the method, and, notably, we observed that this kind of approach can produce small architectures, yet capable of achieving high level of accuracy in a non-trivial biomedical image segmentation task. We do hope that this work can contribute at boosting research for the application of evolutionary methods on image processing, especially while dealing with biomedical tasks.

References

1. Hinton, G.E., Osindero, S., Teh, Y.W.: A fast learning algorithm for deep belief nets. Neural Comput. **18**(7), 1527–1554 (2006)
2. LeCun, Y., Boser, B.E., Denker, J.S. Henderson, D., Howard, R.E., Hubbard, W.E., Jackel, L.D.: Handwritten digit recognition with a back-propagation network. NIPS, pp. 396–404 (1989)
3. Krizhevsky, A., Sutskever, I., Hinton, G.: Imagenet classification with deep convolutional neural networks. Adv. Neural Inf. (2012)
4. Coleman, O.J.: Evolving neural networks for visual processing. Ph.D. thesis, The University of New South Wales, Kensington, Australia (2010)
5. Tan, M., Deklerck, R., Jansen, B., Cornelis, J.: Analysis of a feature-deselective neuroevolution classifier (FD-NEAT) in a computer-aided lung nodule detection system for CT images. In: Proceedings of the 14th Annual Conference Companion on Genetic and Evolutionary Computation, pp. 539–546, ACM (2012)
6. Stanley, K.O., Gauci, J.: Generating large-scale neural networks through discovering geometric regularities. Proc. Genet. Evolutionary Comput. Confrence (GECCO), 2007
7. Miikkulainen, R., Hausknecht, M., Khandelwal, P., Stone, P.: HyperNEAT-GGP: A HyperNEAT-based atari general game player. Genet. Evolutionary Comput. Conference (GECCO), 2012
8. Harguess, J., Verbancsics, P.: Generative neuroEvolution for deep learning. arXiv:1312.5355 (2013)
9. Harguess, J., Verbancsics, P.: Feature learning HyperNEAT: evolving neural networks to extract features for classification of maritime satellite imagery In:. International Conference on Information Processing in Cells and Tissues, Springer International Publishing (2015)
10. Kocmánek, T.: HyperNEAT and Novelty Search for Image Recognition. Diss. Master's thesis, Czech Technical University in Prague (2015)
11. Stanley, K.O., D'Ambrosio, D.B., Gauci, J.: A hypercube-based encoding for evolving large-scale neural networks. Artif. Life **15**(2), 185–212 (2009)
12. Lusk, S.M.: Evolving Neural Networks with Hyperneat and Online Training. Texas State University, Diss (2014)
13. Ganin, Y., Lempitsky, V.: N4-fields: neural network nearest neighbor fields for image transforms. In: Asian Conference on Computer Vision, pp. 536-551 (2014)
14. Marin, D., Aquino, A., Gegundez-Arias, M.E., Bravo, J.M.: A new supervised method for blood vessel segmentation in retinal images by using gray-level and moment invariants-based features. IEEE Trans. Med. Imaging **30**, 146–158 (2011)

Chapter 18
An End-To-End Unsupervised Approach Employing Convolutional Neural Network Autoencoders for Human Fall Detection

Diego Droghini, Daniele Ferretti, Emanuele Principi,
Stefano Squartini and Francesco Piazza

Abstract In the past few years, several works describing systems for the prompt detection of falls have been presented in literature. Many of these systems address the problem of fall detection by using some handcrafted features extracted from the input signals. In the meantime interest in the use of feature learning and deep architectures has been increasing, thus reducing the required engineering effort and the need for prior knowledge. A fall detection method based on a Deep Convolutional Neural Network Autoencoder is presented in this work. This method is trained as a novelty detector through the end-to-end strategy. The classifier distinguishes normal sound events generated by common indoor human activity (i.e. footsteps and speech) and music background from novelty sound events produced by human falls. The performance of the algorithm has been assessed on a corpus of fall events created by the authors. Moreover a comparison was made with two different state-of-art algorithms both based on a One Class Support Vector Machine. The results showed an improvement on performance of about 11% on average.

Keywords Acoustic fall detection · Novelty detection · Autoencoder · Deep neural network

D. Droghini · D. Ferretti (✉) · E. Principi · S. Squartini · F. Piazza
Department of Information Engineering, Università Politecnica delle Marche,
Via Brecce Bianche, 60131 Ancona, Italy
e-mail: d.ferretti@pm.univpm.it

D. Droghini
e-mail: d.droghini@pm.univpm.it

E. Principi
e-mail: e.principi@univpm.it

S. Squartini
e-mail: s.squartini@univpm.it

F. Piazza
e-mail: f.piazza@univpm.it

18.1 Introduction

The Population aging represents a major challenge for the immediate future of both industrialized and developing countries. Estimates show that by 2050 the elderly proportion will tend to double from 11 to 22% [1]. The increase in life expectancy joint to a getting worse ratio between the active and inactive people, will increase the relative socio-economic burden for healthcare and services for elders. The governments are investing in intelligent technologies which help elderly people to live in autonomy in their home, in order to reduce the impact of this demographic change on the society [2]. In this context, an important topic is represented by falls detection. It was observed that about 62% of injury-related hospitalizations for the people over 65 years are the result of a fall. Instead a prompt detection of falls reduces the correlated risks of morbidity and mortality [3].

In the last years, several works have been presented in the literature that describe in-home healthcare monitoring systems and human fall detection techniques [4–7]. The sensors at their basis represent an important discriminative regarding the nature of the different solutions. The sensors employed can be "wearable" if they are worn by the monitored person [8–10] or can be "environmental" (e.g., infrared sensors, pressure, microphones, cameras) if they are arranged in the smart environment [4, 11, 12]. Among the latter are often considered apart the "video-based" sensors, since they refer to the well explored area of computer vision [13].

A second aspect for which the different approaches are distinguished regards the employed algorithm. The "analytical methods" distinguish between fall and non-fall events by applying a threshold directly on the acquired signals or on the features sequences extracted from them [6]. These methods are generally built exploiting some a-priori knowledge to operate in a specific scenario and needs manual tuning of the hyperparameters of the algorithm. For these reasons, the "analytical methods" can hardly perform when the operating conditions and the subjects are variable. In "machine learning" methods, the algorithm learn from the data how to discriminate falls from non-falls [6]. Between them can be distinguished "supervised" and "unsupervised" approaches. The fist require a labelled dataset for training the classifier, while the latter build a normality model considering only the non-fall events. Regardless of the used approach, machine learning tasks require that the inputs are mathematically and computationally convenient to process, so researchers have traditionally relied on a two-stage strategy: some features are extracted from the raw signals of dataset and are then used as input for the successive tasks. The choice and design of the appropriate features requires considerable expertise about the problem and constitutes a significant engineering effort.

In recent years, thanks to the success of deep learning methods have become increasingly popular the feature learning approaches that independently transform the raw data inputs to a representation that can be exploited in machine learning tasks, minimizing the need of prior knowledge of application domain. Furthermore, such approaches are often able to generalize well real-world data compared to traditional hand-crafted features [14], resulting in an increase in performance of classification or regression tasks. The end-to-end learning is a particular example of feature learning, where the entire stack, connecting the input to the desired output, is learned from data [15]. As in feature learning, only the tuning of the model hyperparameters requires some expertise, but even that process can be automated [16].

In this work, an end-to-end acoustic fall detection approach is presented. A deep convolutional neural network autoencoder is trained with the signals, gathered by a Floor Acoustic Sensor, corresponding to sounds that commonly occurring in a home (e.g., voices, footsteps, music, etc.). Since the sound produced by a human fall should be considerably different from the ones used for the training, it will be recognized as "novelty" by the network and classify as Fall. The performance of the algorithm has been evaluated on a corpus created by the authors, which contains human fall events and sounds related to common human activities. In particular the human fall events are simulated by employing the Rescue Randy human mimicking doll [12, 17, 18].

The rest of the paper is organized as follows: Section 18.2 present an overview of some fall detection algorithms related to the proposed approach. In Sect. 18.3 the proposed fall detection algorithm is explained. Section 18.4 describes the experiments conducted and show the resulting performance of the approach. Finally, Sect. 18.5 concludes the paper and presents future developments.

18.2   Related Works

As aforementioned, fall detection approaches can be distinguished based on their sensing technologies and on the algorithm that discriminates falls from non-falls. Many of these approaches have been analyzed in several reviews about this topic [4, 5, 8, 11, 13]. Below, are briefly summarized two works that were chosen as a reference to compare the performance of the proposed approach and then, are mentioned some of works that exploit the potential of end-to-end approach in different tasks respect fall detection.

A major issue in the design of a fall detection system is that it is very challenging to obtain realistic fall sound signatures for training purposes. By the way, Popescu and Mahnot [19] investigate an unsupervised method available when in a binary classification problem there is not data for one of the two classes. They evaluate three different one-class classifier: Gaussian Mixture Models, nearest neighbour and One-Class Support Vector Machine (OCSVM). The dataset used for the experiments is acquired with a single aerial microphone. It comprises falls and non-falls sounds represented by dropping objects, knocking, clapping and phone-calling related. The Mel-Frequency Cepstral Coefficients (MFCCs) are extracted from the dataset. Than each acoustic signal is processed on windows of 1 s.

In a previous works by the authors [20] is presented a solution that combine OCSVM and template-matching classifier to obtain a semi-supervised framework for Fall Detection. The acoustic signals are captured by means of same Floor Acoustic Sensor used in this work and then MFCCs and Gaussian Mean Supervectors (GMSs) are extracted. GMSs are higher level features computed by adapting the means of a Gaussian mixture model (GMM) with maximum a posteriori algorithm (MAP). In the training phase, a large set of audio data is used to model an Universal Background Model (UBM) composed of the GMM extracted by using Expectation Maximization (EM) algorithm [21]. Then, the GMS of each event is calculated by adapting the GMM with the MAP algorithm and concatenating the resulting GMM mean values. In a first stage of the system, abnormal acoustic events are discriminate from normal ones employing the OCSVM classifier. The second stage perform a template-matching to make the final fall/non-fall decision. The used template are composed by those supervector detected as abnormal by the OCSVM and marked as false positives by the user. The performance of the algorithm has been evaluated on a corpus containing sounds of human falls, falling objects, human activities, and music.

Up to the authors' knowledge, the end-to-end strategy has never been applied to a unsupervised fall detection approach with acoustic sensors. However, many works in other research fields can be founds in literature. Dieleman et al. [22] address the content-based music information retrieval tasks, investigating whether it is possible to apply end-to-end feature learning directly to raw audio signals instead that on a spectrogram representation of data. Their convolutional neural networks trained on raw audio signals do not outperform a spectrogram-based approach in the automatic tagging task but are able to autonomously discover frequency decompositions from raw audio, as well as phase and translation-invariant feature representations.

The contribution of this paper is a novel fall detection method which exploits the end-to-end approach, resulting in a reduction of the engineering effort needed to design the system, with respect to those method that used features extracted from input signals. Is also shown that it can achieve good performance, higher than two state-of-the-art systems chosen for a comparison.

18.3   Proposed Approach

In the state of the art the majority of fall detection systems are logically designed as a cascade of elements which perform some sub-task (e.g. feature extraction, modeling, classification, ecc.). Each task is independently developed and generally requires the tuning of a number of hyperparameters using an experimental procedure and some a prior knowledge about the domain of the problem.

The proposed approach, showed in Fig. 18.1, is designed according to the end-to-end paradigm and then, the entire stack, connecting the input to the desired output, is learned from data. End-to-End is a feature learning strategy that, in presence of sufficient training data, may result in better performance than systems based on handcrafted features, since the training procedure automatically select the salient information. Therefore, if were possible to analyze the feature learned by the network with end-2-end strategy, there should be clues about what kind of information is important for a specific task.

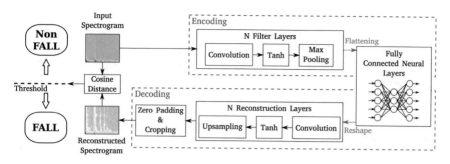

Fig. 18.1 The proposed approach scheme

The system core is a deep convolutional neural network autoencoder. Some exhaustive discussion about this type of network can be easily found in literature [23–25]. The network input consists of the normalized log-power spectrograms of the signals calculated with a STFT on windows of 32 ms and overlapped by 50%. Due to the presence of fully-connect neural layers, the input dimensions must be fixed. After having identified the widest spectrogram extracted from the dataset, the other ones have been extended with some AWGN frames added at the end. Each input consist in a $f \times t$ matrix, where f are the positive points of discrete Fourier transform and t are the number of windows considered in time. The output of the autoencoder are the reconstructed spectrograms. To classify an event, a distance measurement between input and output must be made with some heuristic. If the distance exceeds a certain threshold, automatically defined by the algorithm during the training phase, the system label the output as "Fall" or as "Non Fall" otherwise. In this work the cosine distance has been used:

$$D_C(v, u) = 1 - \frac{u \cdot v}{\|u\| \, \|v\|} = 1 - \frac{\sum_{k=1}^{n} u(k)v(k)}{\sqrt{\sum_{k=1}^{n} u(k)^2} \sqrt{\sum_{k=1}^{n} v(k)^2}} \qquad (18.1)$$

where u and v are the vectors obtained flattening the input and the output spectrograms and n are the length of this vectors. According to the cosine definition, the value of the distance always has a value between -1 and $+1$, where $+1$ indicates two equal vectors while -1 indicates two opposite vectors. The added AWGN part of the spectrums was not considered to calculate the distance. The choice of this heuristic allowed to make distance measurements independents of the size of the initial spectrum. The structure of the autoencoder is not defined a priori, but it is chosen through a phase of cross-validation during which the network parameters are varied with a random search strategy.

18.4 Experiments

In this section are described the composition of the dataset used in this work and the experimental set-up.

18.4.1 Dataset

The employed dataset has been acquired in a rectangular room measuring about 7×2 m using a Presonus AudioBox 44VSL sound card and the FAS positioned on the floor. It is composed of two type of sounds: the first, namely novelty, comprises several human fall sounds that have been simulated by means of Rescue Randy, a human-mimicking doll employed in water rescues. The doll has been dropped from upright position and from a chair, both forward and backward. The drops have been then repeated at three distances from the FAS, i.e., 2, 4 and 6 m, for a total of 44 events, all included in the Human fall class. The second type, i.e, the background, comprises sounds of normal activities (voices, footsteps, etc.) for a total of 593 s, and three musical tracks,[1,2,3] for a total of 1180 s, played from a loudspeaker and acquired back with the FAS. In addition, the signals of the second type have been employed alone and to create noisy versions of human falls occurrences in order to assess the algorithm in presence of interferences. The signals have been acquired with a sampling rate equal to 44.1 kHz and 32 bit depth. Based on previous experience in using the FAS [26], in order to exploit the acoustic characteristics of the sensor, signals have been downsampled to 8 kHz and the resolution has been reduced to 16 bit.

18.4.2 Experimental Setup

Since in this work a novelty approach is presented, the dataset has been divided in two groups: the former composed only of background sounds (i.e human activity sounds and musical background) used for the training; the latter composed of both background sound and novelty sounds, i.e, the human falls, used in development and test phase. In order to assess the classification accuracy in noisy conditions, a second version of human fall sounds were crated in which a musical background was recorded and then digitally added to the fall events.

The input spectrograms of the audio signals has been calculated with a fft point number of 256 and a windows size of 256 samples (32 ms at sample rate of 8000 kHz). The longest spectrogram present in the dataset is composed of 197 frame. Therefore the resulting input matrix dimension $f \times t$ is 129×197. The optimization of the

---

[1] W. A. Mozart, "Piano trio in C major".

[2] Led Zeppelin, "Dazed and confused".

[3] Led Zeppelin, "When the levee breaks".

Table 18.1 Hyper-parameters optimized in the random-search phase, and their range

Parameter	Range	Distribution	Parameter	Range	Distribution
Cnn layer Nr.	[1–3]	Uniform	Batch size	[10–25%]	Log-uniform
Kernel shape	[3 × 3–8 × 8]	Uniform	Max pool shape	[1 × 1–5 × 5]	Uniform
Kernel Nr.	[4–64]	Log-uniform	Max Pool	All[a]-Only end[b]	Uniform
MLP layers Nr.	[1–3]	Uniform	Dropout	[Yes-No]	Uniform
MLP layers dim.	[128–4096]	Log-unifom	Drop rate	[0.5–0.6]	Normal
Stride	[1 × 1–3 × 3]	Uniform	Learning rate	$[10^{-4}$–$10^{-2}]$	Log-unifom

[a] After each Conv. layer
[b] At the end of cnn part

experiment hyper-parameters has been carried out using the random-search technique. Table 18.1 shows the parameters used in the random-search, and their ranges. The parameters of the network architecture are related only to the encoding part of autoencoder since the decoding part is its mirrored version. Instead other parameters, described below, have been set to the same value for all experiments. The activation function for each layer, whether they are convolutional or fully connected, have been set to *tanh*. "Adam" [27] has been used as optimization algorithm for the traing phase. The loss function used was *mlse*. The initialization algorithm for the weight of the autoencoder was Glorot Uniform [28]. The number of epoch has been set to 1000, while the patience, that is the number of epoch without an Auc improvement on a devset to wait before stopping the training phase, has been set to 40.

In order to implements a 4 fold cross-validation, the signals not being part of training-set have been divided in four folds, each composed of 11 human falls and 11 non-falls signals. Then, one fold has been used as validation-set and the remaining three for calculating the performance in test phase. In cross-validation phase the scores have been evaluated in term of AUC. Here also the optimal thresholds have been infer by searching points on ROC curves closest to the $(0, 1)$: $d_{min} = \sqrt{(1 - fpr)^2 + (1 - tpr)^2}$. At the end the final performance has been evaluated in term of $F_1 - Measure$ by mediating the results obtained on individual folds.

The proposed approach has been compared with 2 algorithms both based on OCSVM. In the first algorithm, proposed in a previous work of the authors [20], 13 MFCCs, plus first and second derivatives extracted from a 8 kHz audio signals, were used for training a Universal Background Model (UBM) composed of a Gaussian Mixture Model (GMM) with the Expectation Maximization algorithm [21]. Then, a Gaussian Mean Supervector (GMS) was calculated by adapting the GMM with the MAP algorithm [29] and concatenating the adapted GMM mean values. In the second algorithm, presented in [19], the audio signals are divided in windows of the same lengths, and the related MFCCs are used for training the OCSVM and for classification. Both for comparison purposes and for different composition of the dataset, we have introduced some changes to the original approach: we employee the same MFCCs used in [20]. The window length used for the analysis corresponds to the duration of the shortest event in our dataset, and it is equal to 576 ms (71 frames). Windows are overlapped by 50%, and, as in [19], an event is classified as fall if at least two consecutive frames are classified as novelty by the OCSVM. On both the target algorithms, the grid search procedure has been adopted to find the optimal values ν and γ of the OCSVM and the number of mixture of the GMM-UBM.

18.4.3 Results

The results for both clean and noisy are reported in Fig. 18.2. The comparative algorithms are denoted with "Popescu (2009)" and"OCSVM" respectively, while the proposed approach is named "Autoencoder". It is immediately clear that the proposed approach outperforms the other in both conditions. In fact, in clean condition, it gains about 1% compared to "OCSVM" and about 18.6% compared to "Popescu (2009)". Moreover the proposed algorithm results to be very robust in noisy condition, whereas both other cases get worse a bit. In particular the performance improves by 3.72% with respect to "OCSVM" and by 20.45% with respect to"Popescu (2009)". Clearly the end-to-end method seems insensitive with respect to the corrupted human fall signals when the novelty sounds are dissimilar respect to normality model learned from the background sounds. In Table 18.2 are reported the hyperparameters that have led to the results discussed above.

Furthermore other experiments were made up with a manual tuning of parameters. Particularly have been investigated deeper architectures composed up to 5 convolutional layer. We found that increasing the depth on cnn part (5 layers) with a different

Fig. 18.2 Results in *clean* and *noisy* conditions for the three test cases

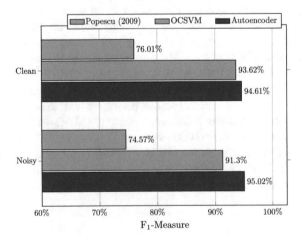

Table 18.2 Best hyper-parameters find in random-search phase for *clean* and *noisy* condition

Parameter	Clean				Noisy			
	Fold1	Fold2	Fold3	Fold4	Fold1	Fold2	Fold3	Fold4
Cnn layer Nr.	3	3	3	2	3	3	3	3
Kernel shape	8×8	7×7	5×5	8×8	8×8	7×7	8×8	8×8
Kernel Nr.	[16, 16, 8]	[32, 16, 16]	[8, 8, 8]	[32, 16]	[32, 32, 8]	[32, 32, 8]	[32, 32, 32]	[8, 8, 8]
Max pool position	Only end	Only end	All	All	Only end	Only end	All	Only end
Max pool shape	5×5	3×3	5×5	4×4	3×3	5×5	5×5	3×3
Stride	3×3	3×3	1×1	3×3	3×3	3×3	1×1	3×3
MLP layers Nr.	2	1	1	1	2	2	1	3
MLP layers dim.	[16, 231]	96	32	32	[48, 153]	[16, 2084]	128	[48, 1952, 1952]
Learning rate($\times 10^4$)	4.89	4.08	15.09	15.44	1.56	4.46	1.01	1.00
Batch size%	11.26	10.81	13.59	21.10	20.06	12.55	13.51	13.13
Drop rate	0.64	0.57	0.53	0.55	0.58	0.53	0.55	0.59

kernel number for each layers of [32, 32, 16, 16, 8] and a kernel dimension for all layers of 4×4, a max pooling after only the first three convolutional layer of 2×2 and two MLP layer of 1024 and 512, leads to considerable improvements. In effect the final $F_1 - Measure$ rise up to 95, 42% in both clean and noisy condition.

18.5 Conclusion

In this paper, the authors proposed an end-to-end approach composed by a deep-convolutional-autoencoder with a downstream threshold classifier, that is a purely unsupervised approach to acoustic fall detection. Our method exploits the reconstruction error of the autoencoder. When a sound that the network has never seen in training phase occurs the reconstruction error raise up allowing the recognition of novelty. The algorithm has been trained with a large corpus of background signals, that is human activity noise and music, and evaluated with human-fall sound and other instances of background sounds. It has been evaluated in two different condition: the first with a clean version of human fall sounds and the second with corrupted version of the same. Moreover a comparison was made with two different algorithms, one proposed in [19] and the other based on OCSVM [20]. The results showed that the proposed solution leads to an average improvement about 20% with respect to the [19] and about 2.3% towards the OCSVM based approach.

In future works this approach will be assessed in a more realistic scenario including, in the test phase, others sound topologies such as falls of different objects, street sounds, etc., that could compromise the classification. In fact those sounds are much more similar to the human falls and could lead to a worse performance. In addition, having regard the results obtained with a manual tuning of parameters, deeper network architectures in combination with Recurrent Neural Network will be thoroughly investigated.

References

1. Organization, W.H., et al.: World health day 2012: ageing and health: toolkit for event organizers (2012)
2. van den Broek, G., Cavallo, F., Wehrmann, C.: AALIANCE Ambient Assisted Living Roadmap, Ambient Intelligence and Smart Environments Series, vol. 6. IOS press, Amsterdam, The Netherlands (2010)
3. Gurley, R.J., Lum, N., Sande, M., Lo, B., Katz, M.H.: Persons found in their homes helpless or dead. New Engl. J. Med. **334**(26), 1710–1716 (1996)
4. Igual, R., Medrano, C., Plaza, I.: Challenges, issues and trends in fall detection systems. BioMedical Engineering Online 12(1) (2013), cited By 111
5. Mubashir, M., Shao, L., Seed, L.: A survey on fall detection: principles and approaches. Neurocomputing **100**, 144–152 (2013)

6. Noury, N., Fleury, A., Rumeau, P., Bourke, A., Laighin, G., Rialle, V., Lundy, J.: Fall detection-principles and methods. In: Engineering in Medicine and Biology Society, 2007. EMBS 2007. 29th Annual International Conference of the IEEE. pp. 1663–1666. IEEE (2007)
7. Principi, E., Fuselli, D., Squartini, S., Bonifazi, M., Piazza, F.: A Speech-Based System for In-Home Emergency Detection and Remote Assistance. In: Proceeding of the 134th International AES Convention. pp. 560–569. Rome, Italy (May 4–7 2013)
8. López-Nava, I.H., Muñoz-Meléndez, A.: Wearable inertial sensors for human motion analysis: a review. IEEE Sens. J. **16**(22), 7821–7834 (2016)
9. Mukhopadhyay, S.C.: Wearable sensors for human activity monitoring: a review. IEEE Sens. J. **15**(3), 1321–1330 (2015)
10. Yang, C.C., Hsu, Y.L.: A review of accelerometry-based wearable motion detectors for physical activity monitoring. Sensors **10**, 7772–7788 (2010)
11. Cippitelli, E., Fioranelli, F., Gambi, E., Spinsante, S.: Radar and rgb-depth sensors for fall detection: a review. IEEE Sens. J. (2017)
12. Zigel, Y., Litvak, D., Gannot, I.: A method for automatic fall detection of elderly people using floor vibrations and sound? proof of concept on human mimicking doll falls. IEEE Trans. Biomed. Eng. **56**(12), 2858–2867 (2009)
13. Erden, F., Velipasalar, S., Alkar, A.Z., Cetin, A.E.: Sensors in assisted living: a survey of signal and image processing methods. IEEE Signal Process. Mag. **33**(2), 36–44 (2016)
14. Principi, E., Squartini, S., Piazza, F.: Power Normalized Cepstral Coefficients based supervectors and i-vectors for small vocabulary speech recognition. In: Proceedings of the International Joint Conference on Neural Networks (IJCNN). pp. 3562–3568. Beijing, China (Jul 6–11 2014)
15. Muller, U., Ben, J., Cosatto, E., Flepp, B., Cun, Y.L.: Off-road obstacle avoidance through end-to-end learning. In: Advances in neural information processing systems. pp. 739–746 (2006)
16. Bergstra, J., Yamins, D., Cox, D.: Making a science of model search: hyperparameter optimization in hundreds of dimensions for vision architectures. In: International Conference on Machine Learning. pp. 115–123 (2013)
17. Alwan, M., Rajendran, P.J., Kell, S., Mack, D., Dalal, S., Wolfe, M., Felder, R.: A smart and passive floor-vibration based fall detector for elderly. Proc. of Inf. Commun. Technol. **1**, 1003–1007 (2006)
18. Werner, F., Diermaier, J., Schmid, S., Panek, P.: Fall detection with distributed floor-mounted accelerometers: An overview of the development and evaluation of a fall detection system within the project eHome. In: Proceedings of the 5th International Conference on Pervasive Computing Technologies for Healthcare and Workshops. pp. 354–361. Dublin, Ireland (May 23–26 2011)
19. Popescu, M., Mahnot, A.: Acoustic fall detection using one-class classifiers. In: Proceeding of the Annual International Conference of the IEEE Engineering in Medicine and Biology Society (EMBC). pp. 3505–3508. Minneapolis, MN, USA (2009)
20. Droghini, D., Ferretti, D., Principi, E., Squartini, S., Piazza, F.: A combined one-class svm and template matching approach for user-aided human fall detection by means of floor acoustic features. Computational Intelligence and Neuroscience 2017 (2017), Article ID 1512670
21. Bilmes, J.A.: A gentle tutorial of the EM algorithm and its application to parameter estimation for Gaussian mixture and hidden Markov models. Technical Reports ICSI-TR-97-021, University of Berkeley (1997)
22. Dieleman, S., Schrauwen, B.: End-to-end learning for music audio. In: Acoustics, Speech and Signal Processing (ICASSP). In: 2014 IEEE International Conference on. pp. 6964–6968. IEEE (2014)
23. Krizhevsky, A., Sutskever, I., Hinton, G.E.: Imagenet classification with deep convolutional neural networks. In: Advances in neural information processing systems. pp. 1097–1105 (2012)
24. Marchi, E., Vesperini, F., Squartini, S., Schuller, B.: Deep recurrent neural network-based autoencoders for acoustic novelty detection. Computational intelligence and neuroscience 2017 (2017)
25. Ng, A.: Sparse autoencoder. CS294A Lecture Notes **72**(2011), 1–19 (2011)

26. Principi, E., Droghini, D., Squartini, S., Olivetti, P., Piazza, F.: Acoustic cues from the floor: a new approach for fall classification. Expert Syst. Appl. **60**, 51–61 (2016)
27. Kingma, D., Ba, J.: Adam: a method for stochastic optimization. arXiv preprint arXiv:1412.6980 (2014)
28. Glorot, X., Bengio, Y.: Understanding the difficulty of training deep feedforward neural networks. Aistats. **9**, 249–256 (2010)
29. Reynolds, D., Quatieri, T.: Speaker verification using adapted gaussian mixture models. Digital Signal Process **10**(1), 19–40 (2000)

Chapter 19
Bot or Not? A Case Study on Bot Recognition from Web Session Logs

Stefano Rovetta, Alberto Cabri, Francesco Masulli and Grażyna Suchacka

Abstract This work reports on a study of web usage logs to verify whether it is possible to achieve good recognition rates in the task of distinguishing between human users and automated bots using computational intelligence techniques. Two problem statements are given, offline (for completed sessions) and on-line (for sequences of individual HTTP requests). The former is solved with several standard computational intelligence tools. For the second, a learning version of Wald's sequential probability ratio test is used.

Keywords Web bot recognition · Classification · Sequential decision

19.1 Introduction

Autonomous Internet agents, or web robots or "bots," are programs that perform a specific action on computers connected in a network, without the intervention of human users. Although statistics are hard to verify, one study [1] estimated that bots account for an average of more than half (51.8%) of the traffic on websites. Among these, again more than half (28.9%) are "bad" bots with malicious goals.

S. Rovetta (✉) · A. Cabri · F. Masulli
DIBRIS – University of Genova, Via Dodecaneso 35, 16146 Genova, Italy
e-mail: stefano.rovetta@unige.it

A. Cabri
e-mail: alberto.cabri@dibris.unige.it

F. Masulli
Sbarro Institute for Cancer Research and Molecular Medicine,
Center for Biotechnology Temple University, Philadelphia (PA), USA
e-mail: francesco.masulli@unige.it

G. Suchacka
Faculty of Mathematics, Physics and Computer Science,
University of Opole, Opole, Poland
e-mail: gsuchacka@uni.opole.pl

© Springer International Publishing AG, part of Springer Nature 2019
A. Esposito et al. (eds.), *Quantifying and Processing Biomedical and Behavioral Signals*, Smart Innovation, Systems and Technologies 103,
https://doi.org/10.1007/978-3-319-95095-2_19

This work reports on the preliminary results of a study of web usage logs of an e-commerce website (which we cannot disclose). The focus is not methodological but applicative. The main research question is whether it is possible to achieve good recognition rates in the task of distinguishing between legitimate, human users and automated bots using computational intelligence techniques rather than hand-engineered filtering criteria.

After discovering that the answer to this question is satisfactorily positive when working on whole-session features, we also tackled the problem of making this decision on-line or *in real time*, namely, while the session is still not complete.

19.2   Web Bots: Types, Goals, and Modes of Operation

Web bots are used to perform a variety of tasks that would be tedious for humans. Although historically the first and primary application is web crawling, many other types of bots exist. Among these we can mention social bots, those that post automatically generated messages on services like Twitter or Telegram; conversational bots [2], automatically replying to questions by human users with varying degrees of complexity, from simply reacting to keywords to engaging in natural language interaction using artificial intelligence; and Wikipedia bots, performing routine maintenance, such as adding templates and replacing text. Bots of these categories are collaborative agents. They usually comply with the directives that website maintainers place in the *robots.txt* file to prevent visits to specific page subsets, have legitimate goals, and their only potential risk is an increase in network traffic, which is usually kept under control by the bots themselves by limiting their own rate of activity.

Along with these "ethical" bots, however, there are other categories of malicious bots. Botnets or bot farms are ensembles of computers running bots used to perform malicious activities from a number of different IP addresses, making it very hard to blacklist them. A botnet is created by infecting computers via malware, which propagates along the usual channels (e.g., mail attachments) and gives control of the infected machine to a central bot controller or "herder" [3].

Botnets can be organised in peer-to-peer or client-server structures. They can use several network protocols; however, due to the overwhelming diffusion of web-based services, HTTP-based bots are the majority [4]. Many bots adopt impersonation tactics, for instance by changing the *user-agent* HTTP/HTTPS header on their own requests to mimic either human users (employing the user-agent string of human-operated web browsers) or "benign" bots like Google or Baidu crawlers [5].

The most common usage of botnets is for performing DDoS (distributed denial of service) attacks with the goal of blocking or gaining access to a particular website or service. However, malicious bots are increasingly also used to gain undue advantage in on-line business. For instance, ticketing bots can be use to buy large quantity of popular event tickets for illegal resale. The so-called click bots are frequently used to simulate a higher number of clicks on web advertising. These so-called "click frauds" operate with either of the following two strategies [6]:

- increasing the number of clicks on one's own website to inflate the website advertising value
- exhausting the budget of one's competing advertisers so that their ads won't appear any more during the current day.

In this work we are interested in filtering bots that are trying to access an e-shop. These bots can serve several different purposes, ranging from click frauds to complete transactions (some bots actually buy items).

19.3 "Bot or Not?": The Bot Recognition Problem

A HTTP session is a sequence of HTTP requests coming from a Web client. Requests are described by records in log files, which collect a number of pieces of information including, among others, the user agent string identifying the client program and the time of issue of the request. Recognizing bots from human users can therefore be formalized as one of the following two problems:

- **Session-based (offline) bot recognition** – Given a set of HTTP request records from a web session, label the session as performed by a bot or by a human.
- **Request-based (on-line) bot recognition** – Given a sequence of HTTP request records from a web session, label the session as performed by a bot or by a human before the sequence ends (if feasible).

The first formulation considers sessions as *sets*, therefore unordered and entirely available at the time of decision making. The other one considers them as *sequences*, with a time ordering and inter-request time intervals. The second problem is often more interesting because it allows one to take actions before the session is over, minimizing the negative effects of the bot visit.

Session identification can also be a problem. Data obtained from web logs do not assign requests to sessions, because HTTP is stateless in itself. Hence, heuristics have to be used. In the present work, requests were simply grouped into one session if generated from the same IP with the same user agent string, assuming a minimum 30-min gap demarcating two consecutive sessions of the same Web client.

19.4 Bot Recognition Techniques

Bot recognition techniques can be classified in various ways. Techniques can address the offline or on-line problem. Depending on the underlying reason for performing the classification, techniques can be tuned to different accuracy/simplicity trade-offs. The sensitivity/specificity balance can also be of concern, because while offline analysis may need to be exhaustive (minimize false negatives), on-line recognition should create as few navigation restrictions as possible to legitimate users (minimize false positives).

Regarding the techniques used [7], syntactical log analysis searches for keywords mainly in user agent strings. This is simple but only possible with a-priori information, which may not be available for new or evolving bots. Traffic pattern analysis looks for known differences in interaction styles between bots and legitimate users. Analytical learning techniques do not search for known patterns, but use statistical/machine learning techniques to learn rules from navigation data. The last category is that of CAPTCHA tests, to recognize users from their ability to solve simple cognitive problems.

Some examples of approaches already experimented are contained in [8], where a traffic pattern analysis was performed, and [9], employing a Bayesian approach.

The present work addresses recognition of bots with the *machine learning* approach to navigational patterns.

19.5 Recognizing Bot Traffic at the Session Level (the Offline Problem)

The offline problem consists in recognizing bots from completed sessions using logged data.

19.5.1 Problem Setting

For the offline task, a set of summary features descriptive of whole sessions were extracted from HTTP logs. The complete list is in Table 19.1. These include overall session information and statistics about requests, including a number for different types, time frequency, and volume transferred. In the feature extraction phase, booleans were encoded as 0/1 while numeric features were individually scaled into [0, 1].

The total available data amounted to 13,587 sessions described by the 22 features listed above, with sessions ranging from as few as 2 requests to some hundreds (sessions containing only one request, albeit present in the original collection, were eliminated from the session set).

Table 19.1 List of features for the offline task

Nr.	Name	Type	Description
Target	is_bot	bool	Whether the session was performed by a bot
1	no_of_pages	int	Number of pages in session
2	no_of_requests	int	Number of requests in session
3	transfer	double	Volume of data transferred in session to the client [KB]
4	purchase	bool	Whether the session ended with a purchase
5	is_admin	bool	Whether the session was performed by the site administrator (also possible bot)
6	duration	int	Duration of the session [s]
7	time_per_page	double	Mean time per page [s] (without taking the last page into consideration)
8	source	bool	Whether a "source" of the session is specified
9	H	int	The number of pages of type "Home" (how many times a home page was opened in session)
10	L	int	The number of pages of type "Login" (operations "Register success" and "Login success")
11	Ship	int	The number of pages of type "Shipping information"
12	S	int	The number of pages of type "Search"
13	B	int	The number of pages of type "Browse"
14	D	int	The number of pages of type "Product details"
15	A	int	The number of pages of type "Add product to cart"
16	I	int	The number of pages of type "Information"
17	E	int	The number of pages of type "Entertainment"
18	empty_ref_pages	[0, 100]	% of pages with empty referrer
19	empty_ref_requests	[0, 100]	% of requests with empty referrer
20	head_requests	[0, 100]	% of requests of type HEAD
21	4xx_requests	[0, 100]	% of erroneous requests (4xx)
22	image_to_page	double	Image-to-page ratio
–	bot_name	string	Also available, but not used: The name of a known bot (or "probably bot" in the case of a probable bot or "–" in the case of a human)

19.5.2   Methodology

In this phase of the work, several standard methods were used to explore the feasibility of offline recognition and assess the quality of the results that can easily be attained with these methods. Two strategies have been experimented with: supervised learning, and unsupervised learning followed by supervised labeling or "calibration".

As representative of supervised methods, two popular classifier models were used: multi-layer perceptron neural networks [10] and support vector machines [11]. The data were randomly split into 70% training set, 15% on-line test set, 15% a posteriori validation set. The test set was used with the multi-layer perceptron for early stopping. Although systematic k-fold cross validation was not used, several random splits were performed to increase the confidence in the obtained results.

The multi-layer perceptron is composed of two layers of computational neurons with non-linear input-output mapping capabilities. Optimisation can be inefficient for very complex problems, but in the present work it proved sufficient to achieve more than satisfactory results. An experimentally-driven model selection step revealed that the performance does not vary much with the number of hidden units above a minimum of two, being very stable up to tens of hidden units, so the selected network size was two hidden units. The model was trained by scaled conjugate gradient with the cross-entropy objective function $-\sum_l (t_l \log y_l - (1 - t_l) \log(1 - y_l))$.

Support vector machines are the default choice for solving two-class classification problems. They were the first successful "kernel method" and build upon statistical learning theory to provide performance guarantees. In this work we employed a RBF (Gaussian) kernel, with the width parameter equal to 0.1.

Regarding unsupervised-plus-labeling methods, we used two central clustering techniques followed by majority labeling of the obtained centroids. The k means method [12] finds crisp clusters, so the labeling step was straightforward. Labelling corresponds to the classic "nearest centroid" classification method.

We also used the *Graded Possibilistic c Means* (GPCM) method [13], a fuzzy clustering method derived from Krishnapuram and Keller's Possibilistic c Means [14]. It allows the detection of outliers and makes learning more robust with respect to location identification (placing centroids), but gives the user more control than its original version. In this case, cluster attributions are fuzzy. Labels for one point are voted by all centroids proportionally to the membership of that point to each, so that the nearest centroids have the most influence. The decision is defuzzified only at the end.

19.5.3 Results

Table 19.2 summarizes the results for multilayer perceptron, SVM, k means with $k = 5, 50, 2011$, GPCM with $k = 5, 50, 2011$. The indicators are accuracy, precision, recall, and F measure. A Receiver's Operating Characteristic curve analysis was not performed because accuracy and F measure are strongly correlated (Pearson correlation coefficient $r = 0.983$ with $P_r = 0.0013$, Spearman rank correlation coefficient $\rho = 0.982$ with $P_\rho = 0.06$), which is an indication that type I and type II errors are effectively balanced.

Table 19.2 Performance indicators for various methods

Method		Accuracy	Precision	Recall	F_1
MLP	#h = 2	0.96	0.91	0.98	0.94
SVM	#sv = 2011	0.98	0.96	0.99	0.98
k means	$k = 5$	0.91	0.86	0.91	0.89
	$k = 50$	0.95	0.90	0.97	0.93
	$k = 2011$	0.98	0.97	0.98	0.98
GPCM	$k = 5$	0.97	0.93	0.99	0.96
	$k = 50$	0.97	0.94	0.99	0.96
	$k = 2011$	0.98	0.95	0.99	0.97

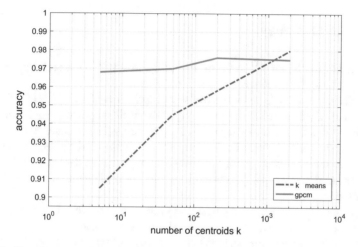

Fig. 19.1 Plot of accuracy versus number of centroids for k means and GPCM

Comparing the results of classification in the fully supervised case with those in the unsupervised + labelling case we can observe that the performance level is surprisingly similar. The quality of errors, however, is more favourable in the supervised case, with a percentage of false positives that is (1) very small in absolute terms, (2) smaller than the corresponding percentage in the unsupervised case, and (3) much smaller than the false negative rate. In the unsupervised + labelling case, the false positive rate is instead larger than the false negative rate. Data are not shown due to space limitations.

The GPCM method yields results which are comparable with those attained by k means, albeit some fraction of a percent inferior in the best case. It should be noted, however, that the decrease in performance with the number of centroids used is much less noticeable in this case. This remark is based on data shown in Fig. 19.1. The GPCM method therefore allows the use of a substantially lower number of centroids, with a modest performance decrease. This makes the method particularly suitable for reverse-engineering the learned classification rule, since the readability of a simple partition is higher compared to that of a complex one.

19.6 Recognizing Bot Traffic at the Request Level (the On-Line Problem)

The on-line problem consists in recognizing bots while their sessions are still active, using request and response data before they are logged.

19.6.1 Methodology

At the time of writing, work on the on-line problem is still in the preliminary phase, but the results look promising.

The on-line problem is an instance of *sequential classification* [15, 16]. The sample is not entirely acquired at the beginning of the decision process, but observations are sampled one at a time, sequentially. A hypothesis (in our case, "the session is performed by a bot") is verified by making observations one at a time. Three outputs are possible for any observation: positive decision, negative decision, and no decision. In the first two cases the sampling stops, in the third one it goes on with a new observation.

If, at observation number t, the posterior probabilities of class 1 (bot) and 0 given the observations $x_1, \ldots, x_t$ are $p_1(t)$ and $p_0(t)$ respectively, and if A, B $(A > B)$ are two thresholds related to the balance between errors of type I (false positive) and II (false negative), then the decision criterion is:

$$\begin{cases} \dfrac{p_1(t)}{p_0(t)} \geq A & \rightarrow \text{Output class 1} \\[2mm] \dfrac{p_1(t)}{p_0(t)} \leq B & \rightarrow \quad \text{Output class 0} \\[2mm] B < \dfrac{p_1(t)}{p_0(t)} < A & \rightarrow \quad \text{Continue sampling} \end{cases} \qquad (19.1)$$

In the original formulation the probabilities are assumed to be known and observations to be mutually independent, so if at step t the class-conditional probabilities of the current observation x_i are $f_1(x_i)$ and $f_0(x_i)$ respectively, we can write the basic *sequential probability ratio*

$$\frac{p_1(t)}{p_0(t)} = \frac{f_1(x_1) f_1(x_2) \cdots f_1(x_t)}{f_0(x_1) f_0(x_2) \cdots f_0(x_t)} = \prod_{i=1}^{t} \frac{f_1(x_i)}{f_0(x_i)} \qquad (19.2)$$

Table 19.3 Performance of on-line classification, individual observations

Net dimensions	Accuracy/sessions (%)	Accuracy/requests (%)
25-13-13-2	80.96	99.07
25-25-25-2	74.46	99.12
25-50-50-2	85.41	99.80
25-50-50-50-2	87.04	99.19

To avoid numerical problems, we logarithmically transformed this expression to work with a sum of differences rather than a product of ratios.

It is interesting to note that the structure so obtained has been conjectured to be biologically plausible after a study on a sequential decision task on monkeys [17]. The probability ratio test in a logarithmic form is a very simple competition between two accumulated sums, so no hypothesis is made on any special neural circuit or computational primitive apart from the canonical ones.

19.6.2   Results

The available on-line training set includes a total of 2,267,504 requests for all the 13,587 sessions. Performing k-fold cross-validation implies training k times a multi-layer perceptron network with $2{,}267{,}504\,(k-1)/k$ requests. Since this part of the research is preliminary, the software to allow these experiments is still in preparation, so results on the whole data set are not available yet. However, the results on different random subsets of 1000 sessions seem to indicate that request features are strongly discriminative.

Table 19.3 compares some network architectures on the tasks of classifying individual requests, compared with the corresponding results for whole sessions (offline task). It is clear that the performance on individual request classification is very good. As a consequence, the sequential solution that uses criterion (19.2) yields almost 100% accuracy. This result is probably overestimated and certainly needs further investigation, but it seems safe to state that the on-line problem can probably be solved to a satisfactory level of accuracy.

19.7   Conclusions

The bot recognition problems considered in this study have been solved with several methods. The results for the offline problem, with summary features, are less favourable than those for the on-line problem. This indicates that point-wise features might carry useful information that is lost in their integration into summary values, e.g., in timing.

In this research only already available techniques have been employed, due to the discovery that some simplifications were possible without sacrificing recognition performance. Future efforts will be directed toward developing a more comprehensive solution for the on-line task, suitable even for the cases where inter-pattern dependencies are not negligible, that is, suitable for sequence or time-series data.

Acknowledgements This work was partially supported by a STSM grant from COST Action IC1406 High-Performance Modeling and Simulation for Big Data Applications (cHiPSet).

References

1. Zeifman, I.: Bot traffic report 2016, https://www.incapsula.com/blog/bot-traffic-report-2016. html, visited on 2017-05-06
2. Microsoft bot framework, https://dev.botframework.com/, visited on 2017-04-20
3. Goodman, N.: A survey of advances in botnet technologies. arXiv preprint arXiv:1702.01132 (2017)
4. Acarali, D., Rajarajan, M., Komninos, N., Herwono, I.: Survey of approaches and features for the identification of http-based botnet traffic. J. Network Comput. Appl. **76**, 1–15 (2016)
5. Bai, Q., Xiong, G., Zhao, Y., He, L.: Analysis and detection of bogus behavior in web crawler measurement. Procedia Comput. Sci. **31**, 1084–1091 (2014)
6. Invalid clicks, https://support.google.com/adwords/answer/42995, visited on 2017-03-21
7. Doran, D., Gokhale, S.S.: Web robot detection techniques: overview and limitations. Data Mining and Knowledge Discovery **22**(1), 183–210 (2011)
8. Suchacka, G.: Analysis of aggregated bot and human traffic on e-commerce site. In: 2014 Federated Conference on Computer Science and Information Systems. pp. 1123–1130 (Sept 2014)
9. Suchacka, G., Sobków, M.: Detection of Internet robots using a Bayesian approach. In: 2015 IEEE 2nd International Conference on Cybernetics (CYBCONF). pp. 365–370 (June 2015)
10. Goodfellow, I., Bengio, Y., Courville, A.: Deep learning. MIT Press, Cambridge (2016)
11. Cortes, C., Vapnik, V.: Support vector networks. Machine Learning **20**, 273–297 (1995)
12. MacQueen, J.: Some methods for classification and analysis of multivariate observations. In: Cam, L.L., Neyman, J. (eds.) Proceedings of the Fifth Berkeley Symposium on Mathematical Statistics and Probability. vol. I, pp. 281–297. University of California (January 1967)
13. Masulli, F., Rovetta, S.: Soft transition from probabilistic to possibilistic fuzzy clustering. IEEE Trans. Fuzzy Syst. **14**(4), 516–527 (2006)
14. Krishnapuram, R., Keller, J.M.: A possibilistic approach to clustering. IEEE Trans. Fuzzy Syst. **1**(2), 98–110 (1993)
15. Ghosh, B.: Sequential Tests of Statistical Hypotheses. Addison-Wesley, Boston (1970)
16. Wald, A.: Sequential tests of statistical hypotheses. The Ann. Mathe. Statist. **16**(2), 117–186 (06 1945)
17. Kira, S., Yang, T., Shadlen, M.N.: A neural implementation of Wald's sequential probability ratio test. Neuron **85**(4), 861–873 (2015)

Chapter 20
A Neural Network to Identify Driving Habits and Compute Car-Sharing Users' Reputation

Maria Nadia Postorino and Giuseppe M. L. Sarnè

Abstract A main question in urban environments is the continuous growth of private mobility with its negative effects such as traffic congestion and pollution. To mitigate them, it is important to promote different forms of mobility among the citizens. Car-sharing systems give users the same flexibility and comfort of private cars but at smaller costs. For this reason, car-sharing has continuously increased its market share although rather slowly. To boost such growth, car-sharing systems needs to increase vehicle fleet, improve company profits and, at the same time, make it more affordable for consumers. In this paper the promotion of car-sharing by reputation is proposed. Neural networks have been used to identify drivers' habits in using car-sharing vehicles. To verify the effectiveness of the proposed approach, some experiments based on real and simulated data were carried out with promising results.

Keywords Driving habits · P2P Car-sharing · Reputation systems · Neural network

20.1 Introduction

The great amount of private cars moving within urban areas is a significant source of traffic congestion and pollution [1–5]. To reduce them, local authorities usually promote transit systems [6–9] also by implementing traffic (e.g., limited traffic zones, interchange areas) [10] and/or monetary policies [11–14]. However, a main question is that cars have more appeal due to comfort, privacy and flexibility with respect to transit [15], which are available only at fixed time and stops, although cars require, in addition to the initial acquisition cost, fixed (e.g., insurance, taxes) and variable (e.g., gas, service) costs to be faced.

M. N. Postorino · G. M. L. Sarnè (✉)
DICEAM, University "Mediterranea" of Reggio Calabria,
Loc. Feo di Vito, 89122 Reggio Calabria, Italy
e-mail: sarne@unirc.it

© Springer International Publishing AG, part of Springer Nature 2019
A. Esposito et al. (eds.), *Quantifying and Processing Biomedical and Behavioral Signals*, Smart Innovation, Systems and Technologies 103,
https://doi.org/10.1007/978-3-319-95095-2_20

In this context, car-sharing systems (CSs) [16] gives the same advantages of a personal car without its disadvantages. More specifically, CS is a membership-based service mainly designed for short trips usually available to the members of a community [17], CSs are characterized by (*i*) users' response and habits [18], (*ii*) cost structure and system organization [19–23], (*iii*) environmental benefits [24, 25]. Currently, three types of CS are diffused, namely:

- *Peer-to-Peer* (P2P): it takes place among private users offering their own cars for money;
- *Business to Consumer* (B2C): it is managed by business companies for profit;
- *Not-For-Profit* (NFP): it is managed by local or social organizations to encourage a sustainable mobility and without profit aims.

The first CS system was realized in Zurich (Switzerland) in 1948. Nowadays, CS activities (mostly B2C) are realized all over the world. Improvements in information and communication technologies are playing a significant role in promoting new CS initiatives as, for instance, allowing private citizens to rent their vehicles when unused [26–29]. However, although CS is growing in popularity, its market share is still marginal with respect to other transport modalities so that suitable policies need to attract new consumers and new investors. Policies have been oriented to discourage the use of private cars by increasing the availability of CS services [18] as for vehicle fleets and urban areas covered by the CS network [30, 31].

As for financial aspects, by analyzing CS costs we can identify the following components:

- *Marketing* - e.g., advertising and other promotional activities;
- *Organization* - e.g., human resources, buildings, parkings;
- *Production* - e.g., management and fleet use.

The attention here is focused on production costs, which also includes those costs due to CS customers' driving habits. For instance, an "aggressive" driving [32] (e.g. speeding up or hard braking) impacts directly on the service costs and indirectly on the profits (given the required vehicle stopping time for maintenance). Such costs may be significant and their reduction could release resources for promoting CS by:

- offering lower fares to users [33, 34];
- increasing CS company profits and attracting new investors/companies or individual car owners in the CS business.

To reach these targets it is crucial to support virtuosos changes of the driving habits in using CS services, which should encourage also more individual owners to share their personal cars when unused. In similar contexts, an approach often exploited is the use of a reputation system. The reputation scores computed for each CS customer on the basis of his/her driving behavior can assign lower fares to better drivers as well as deny the CS service access to the worst drivers. In this scenario, progresses made in different technological fields as computer science, signal processing, communications allow monitoring drivers in an easy way. This paper investigates on supporting

CS activities by a distributed reputation system. Such system classifies CS users based on an artificial neural network that identifies the different driving styles.

In the following, Sect. 20.2 provides an overview of the proposed monitoring system, while the reputation system is described in Sect. 20.3. The results of some experiments carried out to verify the effectiveness of the proposed approach are presented in Sect. 20.4. Finally, in Sect. 20.5 some conclusions are drawn.

20.2   The Proposed Monitoring System

Costs due to management (e.g., buying or renting the fleet, insurances, taxes, information and communication technologies) and the use of the fleet (e.g., maintenance, cleaning) contribute to CS *Production costs* [20]. In particular, inadequate driving habits can lead to accidents or additional time spent in service and maintenance activities, which imply higher costs and also lower profits for the necessary downtime of CS vehicles [35, 36]. Moreover, inappropriate driving behaviors impact negatively on the P2P CS by discouraging private users to share their personal vehicles.

Therefore, encouraging more appropriate driving behaviors will contribute to save money by reducing the CS production costs. In such a context smart CS policies could return a share of saved money to the best drivers by reducing their fares. As a result, all the CS actors could take advantage from good driving practices and a virtuosos process might be realized.

Recent advancements in computer science, control, signal and communication processes allow monitoring vehicles, which currently are equipped with about 120 different types of sensors [37, 38]. The analysis of the information returned by sensors can be exploited to capture driving habits (other than to support new forms of P2P CS [26–28]) and, consequently, classify drivers and assist them in improving their driving habits.

Based on the drivers' classification, it is possible to compute their reputation to set personalized CS fares (e.g. greater or lower discount on the standard price on the basis of the reputation score) or to deny the access to the CS service. To this purpose, a distributed software architecture is proposed, suitable for all the CS systems (i.e., P2P, B2C and NFP), which consists of: (*i*) software entities hosted by each car and (*ii*) a central software unit managing the proposed reputation system.

More in detail, each CS user is monitored by a specific software component [39], plugged into each CS vehicle, exploiting the data gathered by on board sensors. This data analysis is performed in real time for classifying user's driving actions for both (*i*) allowing improvements in his/her driving behavior; (*ii*) computing a score (i.e., feedback) about his/her driving style referred to the current CS session.

As for (*i*), note that, on an increasing number of vehicles drivers already receive some information from their dashboards, which are addressed to make them aware of some specific aspects of their driving as, for instance, the fuel consumption [40]. The other main target of the analysis is the computation of a score, S, for each user's CS session on the basis of his/her driving actions. All the feedbacks computed locally are collected by a central software unit that uses them to update users' global scores, which represent their reputation R (see next section) in using CS services [41].[1] Based on such global scores, a user might either not be allowed to access the CS service or enjoy reduced CS fares.

To this end, let $s_{i,j}$ be the CS service referred to the CS vehicle v_i driven by user u_j. Moreover, let $S_{i,j}$ be the score of u_j, ranging in $[0, 1] \subset \mathbb{R}$, computed by using sensors data of user's driving actions. More specifically, $S_{i,j} = 0$ is the minimum appreciation for u_i with respect to the CS session $s_{i,j}$, while $S_{i,j} = 1$ denotes the maximum one.

More formally, the score $S_{i,j}$ is assumed as:

$$S_{i,j} = \mathcal{F}(s_{i,1}, \ldots, s_{i,n}) \tag{20.1}$$

$S_{i,j}$ depends on data $s_{i,1}, \cdots, s_{i,n}$ provided by the n sensors equipping the vehicle v_i, $\mathcal{F}$ being the adopted function. Different artificial intelligence techniques (e.g., fuzzy-logic [42], artificial neural networks [43], reinforcement learning [44]) can be used to compute $S_{i,j}$ depending on the number of sensors and the nature of provided data. The choice of the best $\mathcal{F}$ to compute a suitable value of $S_{i,j}$ is not a trivial question.

Finally, there could be different modalities to compute S (i.e., by using different functions $\mathcal{F}$) and the necessity of making coherent the timestamp sequence (see Sect. 20.4).

20.3   The Reputation System

The reputation of a person or a thing within a community can be assumed as a collective trustworthiness measure obtained by referrals or ratings given by the other members of the same community that take into account their past interactions [45–47]. Moreover, a good reputation system has to fulfill the following requirements [41];

– reputation scores are time persistent;
– reputation scores are spread within the involved community;
– reputation scores drive decisional processes about future interactions.

To realize the targets described in Sect. 20.2, we conceived a *Reputation System* (RS) for supporting CS actors.[2] The RS exploits the scores computed for each user to represent the CS user' driving habits by means of a synthetic measure. More in detail, when the service s on the vehicle v_i made by the user u_j ends, then a score

---

[1] Note that some events, e.g. car body damages or interior cleaning, cannot be evaluated automatically and require a human intervention to be taken into account in the global reputation score (see Sect. 20.3).

[2] To preserve the user's privacy, the RS associates each user with a unique ID to which each his/her data is referred. Driver and car-owner identities are mutually disclosed only when the CS service is agreed.

$S_{i,j}$ is locally computed on the basis of sensor data and sent to a central software unit (CSU) that updates the global reputation score $R_j \in [0, 1] \subset \mathbb{R}$ of u_j. When the score $S_{i,j}$ for the h-th CS service s consumed by u_j is sent to the CSU, then R_j is updated as follows:

$$R_j^h = (a \cdot F_{i,j} + (1-a) \cdot R_j^{h-1}) \cdot P_j^s \tag{20.2}$$

where:

- a is a system parameter ranging in $[0, 1] \subset \mathbb{R}$, used to tune the sensitivity of the reputation system by weighting the relevance of $F_{i,j}$ (see below).
- $F_{i,j}$ is the contribution to the reputation computed as $F_{i,j} = S_{i,j} \cdot C_{i,j} \cdot B_i$, where:

  • $S_{i,j}$ is the score computed for the CS session carried out by u_j on the vehicle v_i.
  • $C_{i,j}^s$ is a parameter ranging in $[0, 1] \subset \mathbb{R}$ and referred to the cost $c_{i,j}^s$ of the CS service s. More formally, $C_{i,j}^s$ is computed as:

$$C_{i,j}^s = \begin{cases} c_{i,j}^s / C_{Max} & if \;\; c_{i,j}^s < C_{Max} \\ 1 & if \;\; c_{i,j}^s \geq C_{Max} \end{cases} \tag{20.3}$$

  where C_{Max} is a system parameter giving the maximum cost for a CS service after which $C_{i,j}^s$ is assumed to be saturated. $C_{i,j}^s$ should avoid users alternate behaviors addressed to gain reputation on cheap CS services and spend it in expensive CS services by keeping unsuitable driving habits. However, for NFP CS services $C_{i,j}^s$ is fixed to 1.
  • B_j is a system parameter ranging in $[0, 1] \subset \mathbb{R}$. It gives the reputation system a uniform metric by taking into account those characteristics of the CS services not intrinsically considered by the parameter $C_{i,j}^S$ and mainly due to the adoption of different policies by CS companies.

- P_j^s is a penalization coefficient for u_j ranging in $[0, 1] \subset \mathbb{R}$ considering accidents, damages, interior cleaning, penalties, which cannot be processed automatically and require the human participation. B_j^s is set to 1 for default, which corresponds to no penalization for u_j.

On the basis of the global reputation score, different policies can be adopted as, for example, personalized fares or denied access to CS services for users characterized by a low global reputation score.

20.4 Experiments

In this section, the results of some preliminary experiments addressed to verify the effectiveness of the discussed approach for promoting CS activities are presented. More specifically, based on an open database, we studied the nature of the users' driving actions in order to compute locally their driving scores. Then, we tested the effectiveness of the proposed reputation system.

Table 20.1 The adopted aggressive/non aggressive driving actions ratio

Driving category	Aggressive/non aggressive ratio
Very soft	1:0 (i.e., only not aggressive actions)
Soft	From 4:1 to 2:1
Neutral	1:1
Aggresive	From 1:4 to 1:2
Very aggresive	0:1 (i.e., only aggressive actions)

20.4.1 Driving Data

Two experiments have been carried out by exploiting the data publicly available on the OpenXC repository [48]. OpenXC stores data collected by probe-cars and organized in files. Files refer to both different scenarios and driving styles and data are anonymous (i.e., any information useful to associate the tuple to drivers is provided).

More in detail, each OpenXC file consists of tuples with three pairs of data ("name":"string", "value":integer, "timestamp":time). For each pair, the first data identifies the nature of the information and the second data is the associated value. The three pairs identify the type of the gathered information, its measured value and the time of its acquisition respectively. An example of OpenXc tuple is reported below:

{"name" : "accelerator_pedal_position", "value" : 3, "timestamp" : 1483759322.457322}

To perform the experiments we used the OpenXC real data grouped according to driving profiles as in Table 20.1. Each driver profile was associated with a driving style, respectively named *very soft*, *soft*, *neutral*, *aggressive* and *very aggressive*, each one denoting a different level of "aggressive" driving[3] (e.g., hard acceleration, hard braking and so on) on the basis of the number of aggressive actions included in each driving track. The adopted aggressive/non aggressive driving actions ratio are reported in Table 20.1.

To obtain suitable patterns for the following step (see Sect. 20.4.2), we generated 50 drivers' profile, 10 for each driving style. Then, for each driver we generated 20 driving tracks by exploiting the OpenXC data accordingly to his/her driving style. Each driving track stores more driving sequences fitting the assigned driver's profile, where each driving track sequence consists of at least 3 tuples belonging to the same driving style (e.g., "aggressive" or "not aggressive"). Moreover, for each driving track a simple timestamps lineup was needed. Note that this task does not modify the original time sequences and, therefore, it does not affect the obtained experimental

---

[3]Note that the OpenXC repository separates data specifically coming from "aggressive" driving sessions.

results. Finally, the score computation (see Sect. 20.2) is based on the ratio between the numbers of aggressive and all the actions of an examined driving track.

20.4.2   Identification of the Driving Actions

In the proposed system, the right association of each driving action with a driving style plays a fundamental role for computing the drivers' score and then their global reputation score (see Sects. 20.2 and 20.3). To perform this recognition process we exploited an Artificial Neural Network (ANN), a tool able to deal with problems denoted by uncertainty often used in transportation research [49, 50]. In particular, the exploited ANN architecture, topology and learning strategy was identified on the basis of some preliminary tests.

The training ANN pattern is made by 9 input data extracted by three consecutive tuples (identified as "previous","target" and "following"). Each tuple provides three input data (i.e., an integer number coding the attribute "name", the associated value and the time interval corresponding to the action codified by the previous tuple, or 0 if it is the first tuple). The output is a unique real value ranging in [0, 1], where 0/1 is its minimum/maximum aggressiveness degree. By means of a back-propagation (BP) algorithm [43], we trained a three-layer ANN having 9, 120 and 1 nodes for the input, hidden and output layers and respectively hyperbolic and sigmoid activation functions for the neurones of the hidden and output layers. The results obtained by this ANN on unknown patterns in recognizing an aggressive driving action are reported in Table 20.2. These results can be considered satisfactory given the nature of the exploited OpenXC data.

The second experiment tested the effectiveness of the reputation system by simulating 1000 drivers randomly associated with the driving styles of Table 20.1 and by assuming a uniform driving behavior for all the simulation. Moreover, the drivers received a starting global reputation score of 0.5, the system parameters α, P and B were respectively set to 0.15, 1 and 1, while scores and costs were randomly generated coherently with the drivers' habits. For each *epoch* only 20% of the overall number of drivers is randomly selected. Results depicted in Fig. 20.1 shows that the accuracy of the proposed reputation system to recognize the driver nature is about 90% less than 80 epochs.

Note that all the *normal* drivers are recognized thanks to the initial global reputation score of 0.5 and this result does not change along all the simulation.

Table 20.2 ANN accuracy in recognizing aggressive driving actions

Driving style	Very soft	Soft	Neutral	Aggresive	Very aggresive
Percentage	0.93	0.82	0.81	0.79	0.79

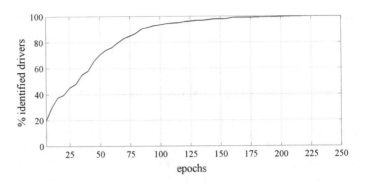

Fig. 20.1 Percentage of drivers correctly identified

20.5 Conclusions

CS can support a more efficient mobility and contribute to improve traffic congestion and pollution impacts in urban contexts. To increase the CS convenience for all the CS actors, in this paper we investigated on a reputation system suitably designed to promote the CS development.

The focus of this proposal is the identification of the drivers' habits which lead to the computation of their global reputation scores. To this purpose, we took advantage from the use of an ANN to identify the driving habits of CS users. The results of some preliminary experiments are promising.

As future works, more complete experimental sessions with a wider set of data will be studied.

Acknowledgements This work has been supported by the *Networks and Complex Systems* Laboratory - Department DICEAM - University Mediterranea of Reggio Calabria.

References

1. Gurjar, B.R., Molina, L.T., Ojha, C.S.: Air pollution: health and environmental impacts. CRC press (2010)
2. Fontes, T., Pereira, S., Fernandes, P., Bandeira, J., Coelho, M.: How to combine different microsimulation tools to assess the environmental impacts of road traffic? lessons and directions. Transportation Research Part D: Transport and Environment **34**, 293–306 (2015)
3. Cascetta, E., Postorino, M.N.: Fixed point approaches to the estimation of o/d matrices using traffic counts on congested networks. Transportation science **35**(2), 134–147 (2001)
4. Seign, R., Schüßler, M., Bogenberger, K.: Enabling sustainable transportation: The model-based determination of business/operating areas of free-floating carsharing systems. Research in Transportation Economics **51**, 104–114 (2015)
5. Velazquez, L., Munguia, N.E., Will, M., Zavala, A.G., Verdugo, S.P., Delakowitz, B., Giannetti, B.: Sustainable transportation strategies for decoupling road vehicle transport and carbon

dioxide emissions. Management of Environmental Quality: An International Journal **26**(3), 373–388 (2015)

6. Postorino, M., Musolino, G., Velonà, P.: Evaluation of o/d trip matrices by traffic counts in transit systems. In: Schedule-Based Dynamic Transit Modeling: theory and applications. Springer (2004) 197–216

7. Harford, J.D.: Congestion, pollution, and benefit-to-cost ratios of us public transit systems. Transportation Research Part D: Transport and Environment **11**(1), 45–58 (2006)

8. Postorino, M., Fedele, V.: The analytic hierarchy process to evaluate the quality of service in transit systems. WIT Transactions on The Built Environment **89** (2006)

9. Dittmar, H., Ohland, G.: The new transit town: best practices in transit-oriented development. Island Press (2012)

10. Hounsell, N., Shrestha, B., Piao, J., McDonald, M.: Review of urban traffic management and the impacts of new vehicle technologies. IET ITS **3**(4), 419–428 (2009)

11. Postorino, M.N.: A comparative analysis of different specifications of modal choice models in an urban area. European journal of operational research **71**(2), 288–302 (1993)

12. Anas, A., Lindsey, R.: Reducing urban road transportation externalities: Road pricing in theory and in practice. Review of Environmental Economics and Policy (2011) req019

13. de Palma, A., Lindsey, R.: Traffic congestion pricing methodologies and technologies. Transportation Research Part C: Emerging Tech. **19**(6), 1377–1399 (2011)

14. Gillen, D.: The role of intelligent transportation systems in implementing road pricing for congestion management1. The Implementation and Effectiveness of Transport Demand Management Measures: An International Perspective **1992** (2016)

15. Katzev, R.: Car sharing: A new approach to urban transportation problems. Analyses of Social Issues and Public Policy **3**(1), 65–86 (2003)

16. Agyeman, J., McLaren, D., Schaefer-Borrego, A.: Sharing cities. Friends of the Earth Briefing (2013) 1–32

17. Sinclair, C.: Codes of ethics and standards of practice. (1993, available at www.carsharing.org)

18. Kopp, J., Gerike, R., Axhausen, K.W.: Do sharing people behave differently? an empirical evaluation of the distinctive mobility patterns of free-floating car-sharing members. Transportation **42**(3), 449–469 (2015)

19. Fellows, N., Pitfield, D.: An economic and operational evaluation of urban car-sharing. Transportation Research D: Transport and Environment **5**(1), 1–10 (2000)

20. Nair, R., Miller-Hooks, E.: Fleet management for vehicle sharing operations. Transportation Science **45**(4), 524–540 (2011)

21. Shaheen, S.A., Cohen, A.P.: Carsharing and personal vehicle services: worldwide market developments and emerging trends. International Journal of Sustainable Transportation **7**(1), 5–34 (2013)

22. Ciari, F., Balac, M., Balmer, M.: Modelling the effect of different pricing schemes on free-floating carsharing travel demand: a test case for zurich, switzerland. Transportation **42**(3), 413–433 (2015)

23. Waserhole, A., Jost, V.: Pricing in vehicle sharing systems: Optimization in queuing networks with product forms. EURO Journal on Transportation and Logistics **5**(3), 293–320 (2016)

24. Rabbitt, N., Ghosh, B.: Economic and environmental impacts of organised car sharing services: A case study of ireland. Research in Transportation Economics **57**, 3–12 (2016)

25. Nijland, H., van Meerkerk, J.: Mobility and environmental impacts of car sharing in the netherlands. Environmental Innovation and Societal Transitions (2017)

26. Car2Share: https://www.car2share.com. (2017)

27. Drife: http://dryfe.it. (2017)

28. Getaround: https://www.getaround.com/san-francisco. (2017)

29. Teubner, T.: Thoughts on the sharing economy. Proceedings of the International Conference on e-Commerce. **11**, 322–326 (2014)

30. Nourinejad, M., Roorda, M.J.: Carsharing operations policies: a comparison between one-way and two-way systems. Transportation **42**(3), 497–518 (2015)

31. Uesugi, K., Mukai, N., Watanabe, T.: Optimization of vehicle assignment for car sharing system. In: International Conference on Knowledge-Based and Intelligent Information and Engineering Systems, Springer (2007) 1105–1111
32. Meseguer, J.E., Toh, C., Calafate, C.T., Cano, J.C., Manzoni, P.: Drivingstyles: A mobile platform for driving styles and fuel consumption characterization. arXiv preprint arXiv:1611.09065 (2016)
33. Quek, Z.F., Ng, E.: Driver identification by driving style. Technical report, Technical report, technical report in CS 229 Project, Stanford university (2013)
34. Henten, A.H., Windekilde, I.M.: Transaction costs and the sharing economy. info **18**(1) (2016) 1–15
35. Zvirin, Y., Tartakovsky, L., Aronov, B., Parent, M.: Modeling vehicle performance for sustainable transport. In: Proc. 17th Int. Symp. on Transport and Air Pollution, Toulouse, Actes INRETS. Number 122 (2009) 17–24
36. Grant-Muller, S., Usher, M.: Intelligent transport systems: The propensity for environmental and economic benefits. Technological Forecasting and Social Change **82**, 149–166 (2014)
37. Baltusis, P.: On board vehicle diagnostics. Technical report, SAE (2004)
38. Telematics: http://www.com/board-diagnostics-future-vehicle-analysis. (2017)
39. Postorino, M.N., Sarné, G.M.L.: Agents meet traffic simulation, control and management: A review of selected recent contributions. In: Proceedings of the 17th Workshop "from Objects to Agents", WOA 2016. Volume 1664 of CEUR Workshop Proceedings., CEUR-WS.org (2016)
40. Hari, D., Brace, C.J., Vagg, C., Poxon, J., Ash, L.: Analysis of a driver behaviour improvement tool to reduce fuel consumption. In: Connected Vehicles and Expo (ICCVE), 2012 International Conference on, IEEE (2012) 208–213
41. Resnick, P., Kuwabara, K., Zeckhauser, R., Friedman, E.: Reputation systems. Communications of the ACM **43**(12), 45–48 (2000)
42. Gottwald, S.: Fuzzy sets and fuzzy logic: The foundations of application from a mathematical point of view. Springer-Verlag (2013)
43. Haykin, S.: Neural Networks - A Comprensive Foundation. Macmillan College Publishing Company, New York (2000)
44. Wiering, M., Van Otterlo, M.: Reinforcement learning. Adaptation, Learning, and Optimization **12** (2012)
45. Jøsang, A., Ismail, R., Boyd, C.: A survey of trust and reputation systems for online service provision. Decision support systems **43**(2), 618–644 (2007)
46. Rosaci, D., Sarné, G.M.L., Garruzzo, S.: TRR: An integrated reliability-reputation model for agent societies. In: Proceedings of the 12th Workshop from "Objects to Agents", WOA 2014. Vol. 741 of CEUR Workshop Proc., CEUR-WS.org (2011)
47. Postorino, M.N., Sarné, G.M.L.: An agent-based sensor grid to monitor urban traffic. In: Proceedings of the 15th Workshop from "Objects to Agents", WOA 2014. Volume 1260 of CEUR Workshop Proceedings., CEUR-WS.org (2014)
48. Openxcplatform: http://openxcplatform.com/resources/traces.html. (2017)
49. Postorino, M.N., Sarné, G.M.L.: A neural network hybrid recommender system. In: Proceedings of the 2011 conference on neural Nets WIRN10. (2011) 180–187
50. Postorino, M.N., Versaci, M.: A neuro-fuzzy approach to simulate the user mode choice behaviour in a travel decision framework. International Journal of Modelling and Simulation **28**(1), 64–71 (2008)

Part III
Neural Networks and Pattern Recognition in Medicine

Chapter 21
Unsupervised Gene Identification in Colorectal Cancer

P. Barbiero, A. Bertotti, G. Ciravegna, G. Cirrincione, Eros Pasero and E. Piccolo

Abstract Cancer is a large family of genetic diseases that involve abnormal cell growth. Genetic mutations can vary from one patient to another. Therefore, personalized precision is required to increase the reliability of prognostic predictions and the benefit of therapeutic decisions. The most important issues concerning gene analysis are strong noise, high dimensionality and minor differences between observations. Therefore, it has been chosen an unsupervised approach in order to bypass the high dimensionality issue using parallel coordinates and a scoring algorithm of features based on their clustering ability. Traditional methods of dimensionality reduction and projection are here used on subset features with high discriminant power in order to better analyze the data manifold and select the more meaningful genes. Previous studies show that mutations of genes NRAS, KRAS and BRAF lead to a dramatic decrease in therapeutic effectiveness. The following analysis tries to explore in an

---

P. Barbiero (✉) · G. Ciravegna · E. Piccolo
Department of Control and Computer Engineering, Politecnico di Torino, Turin, Italy
e-mail: pietro.barbiero@studenti.polito.it

G. Ciravegna
e-mail: gabriele.ciravegna@studenti.polito.it

E. Piccolo
e-mail: elio.piccolo@polito.it

A. Bertotti
Dipartimento di Oncologia, Università degli studi di Torino, Candiolo Cancer Institute - FPO, IRCCS, Candiolo, Italy
e-mail: andrea.bertotti@ircc.it

G. Cirrincione
Laboratory of LTI, University of Picardie Jules Verne, Amiens, France
e-mail: exin@u-picardie.fr

G. Cirrincione
University of South Pacific, Suva, Fiji

E. Pasero
Department of Electronics and Telecommunications, Politecnico di Torino, Turin, Italy
e-mail: eros.pasero@polito.it

© Springer International Publishing AG, part of Springer Nature 2019
A. Esposito et al. (eds.), *Quantifying and Processing Biomedical and Behavioral Signals*, Smart Innovation, Systems and Technologies 103,
https://doi.org/10.1007/978-3-319-95095-2_21

unconventional way gene expressions over tissues which are wild type regarding to these genes.

Keywords Biplot · Calinski-Harabasz index · Colorectal cancer · Curvilinear component analysis · Parallel coordinates · Principal component analysis Xenograft

21.1 Introduction

Cancer is a heterogeneous disease and personalized precision is required to increase the reliability of prognostic predictions and the benefit of therapeutic decisions [1]. In recent years, Patient-Derived Xenografts (PDXs) have emerged as powerful tools for biomarker discovery and drug development in oncology [2–4]. PDXs are obtained by propagating surgically derived tumour specimens in immunocompromised mice and retain the idiosyncratic characteristics of different tumours from different patients. Hence, they can effectively recapitulate the intra- and inter-tumour heterogeneity characterizing human cancer in patients. Building on these premises, PDXs are extensively exploited to conduct large-scale preclinical analyses for drawing statistically robust correlations between genetic or functional traits and sensitivity to anti-cancer drugs. In this context, during the last decade we have been assembling the largest collection of PDXs from metastatic colorectal cancer (mCRC) available worldwide in an academic environment. Such resource has been widely characterized at the molecular level [5–7] and has been leveraged to reliably anticipate clinical findings [8] with major therapeutic implications. Here we propose to exploit available transcriptional data obtained from mCRC PDXs through the Illumina beadarray technology [9] to identify and validate novel algorithms for prediction of drug sensitivity in PDXs and humans. All tissues, considered here, are wild type concerning genes NRAS, KRAS and BRAF. Previous studies show that mutations of these genes lead to a dramatic decrease in Cetuximab effectiveness.

21.2 The High Dimensional Problem

The dataset stemming from the DNA microarrays is composed of the expression of 15,396 genes in 203 mCRC murine tissues. The analysis has been done by considering the genes as variables. This is a very challenging problem because of analyzing very high dimensional data by means of a small training set. Data are preprocessed by the logarithmic normalization and the z-score technique (column statistical scaling) in order to work on the same range amplifying small distances. Since the dataset has a strong noise component and the process is very complex, classical methods of feature selection and dimensionality reduction might provide biased results. Therefore, it has been chosen an unsupervised path in order to bypass the high dimensionality issue through a scoring algorithm of features based on their clustering ability. Traditional methods of dimensionality reduction and projection are here used on subset features with high discriminant power in order to better analyze the data manifold.

21.3   Hierarchical Clustering

Initially, tissues have been grouped considering all the available genes (features) with the Unweighted Pair Group Method with Arithmetic Mean (UPGMA), an agglomerative hierarchical clustering approach [10]. This bottom-up approach finds and merges the nearest pair of clusters r and s according to their mutual average distance:

$$d(r, s) = \frac{1}{n_r n_s} \sum_{i=1}^{n_r} \sum_{j=1}^{n_s} d\left(x_{ri}, x_{sj}\right)$$

The average distance algorithm has been chosen because it is robust against noise and outliers. Among different metrics, the Minkowski distance of order $p = 1$ has been found to have the highest cophenetic correlation coefficient (here 0.8392), therefore it is used in the following analyzes [11]. However, this approach has not provided satisfactory results; indeed, it was not able to find meaningful groups. For this reason, tissues have been processed using a different procedure based on the Ward's minimum variance method [12]. This algorithm finds and merges the pair of clusters that leads to minimum increase in the total within-cluster variance after merging. The within-cluster variance increment due to the merging of r and s is proportional to the distances of the resulting cluster objects from the resulting cluster centroid:

$$d(r, s) = \sqrt{\frac{2n_r n_s}{(n_r + n_s)}} \|x_r - x_s\|_2$$

Differently from the first approach, the clustering algorithm is applied using one feature at a time, determining a one-dimensional clustering which yields the individual ability of discrimination of genes. This property is evaluated using the Calinski-Harabasz index (also called Variance Ratio Criterion, VRC [13]):

$$VRC_k = \frac{SS_B(N - k)}{SS_W(k - 1)}$$

where N is the number of samples, k is the number of clusters, SS_B is the between cluster variance and SS_W is the within cluster variance.

Well defined clusters tend to have a high VRC. Therefore, genes are ranked according to this index. By defining a threshold in advance, several genes can be extracted. In other words, genes are selected according to their ability to discriminate tissues:

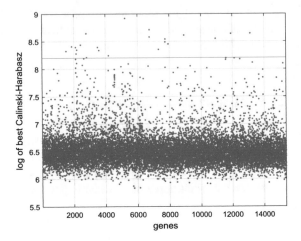

Fig. 21.1 Cluster quality evaluation. The red line is the threshold

this is estimated by checking the best possible separation (with regard to several choices of the number of clusters by means of the parameter k) in terms of quality of the groupings by using an index (see Fig. 21.1).

21.4 Parallel Coordinate Plot

Parallel coordinates are a powerful way of visualizing high-dimensional data. This kind of data visualization was invented during the 19th century and sharpened by Wegman in 1990 [14]. A point in n-dimensional space is represented as a polyline with vertices on equally spaced parallel axes each of one representing a feature; the position of the vertex on the i-th axis corresponds to the i-th coordinate of the point.

This plot is used in order to understand deeply the gene capacity of discrimination, by visualizing the distribution of the murine tissues (colored polylines) along all the dimensions (genes) represented as parallel vertical axes. In this figure blue lines represent worsening cancers, red lines stable cancers and yellow lines regressive cancers, respectively. The intersections of the polylines with the vertical axes show there are some genes highly discriminating the three colors, which means that some mice have particular expression levels for some genes. Also, the colored grouping of polylines show coherency, which means there is discrimination for tissues. Hence, these genes can be used as markers of CRC subtypes. This tool has been used as a visualization tool for the validation of the previous gene selection, based on the cluster quality evaluation. This technique has confirmed the selection of 22 genes (see Fig. 21.2). This allows the study of the data manifold representing the tissues in a lower dimensional space, just alleviating the problem of the curse of dimensionality.

Fig. 21.2 Parallel
coordinate plot

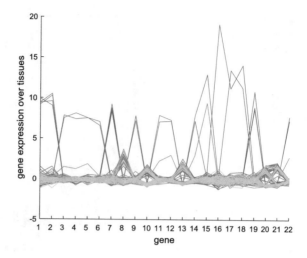

21.5   Analysis of Data Manifold

The reduced submatrix is composed of 203 rows, i.e. the tissue values (samples) and 22 columns, i.e. the selected genes (features). In order to check the intrinsic dimensionality and the linearity of the data manifold, the Principal Component Analysis (PCA [15]) has been performed. The plot of the variance explained by the principal components shows an intrinsic dimensionality of about 5 (see Fig. 21.3), corresponding to 90% of variance explained. This result suggests that tissues belonging to the 22-dimensional spacelay on a 5-dimensional hyperplane. The remaining 10% can be justified by either noise or small departure from linearity, that is nonlinearity only on a large scale, but not locally. In order to confirm this hypothesis, the Curvilinear Component Analysis (CCA [16]) has been used. CCA is a neural technique for dimensionality reduction which projects points by preserving as many distances as possible in the input space. However, CCA is here used not for the exploitation of the projection, but for the information that can be derived from its dy-dx diagram (see Fig. 21.4). This plot represents distances between pairs of points in the input space (the dx value) and in the reduced (latent) space (the dy value) as a pair (dy, dx). If a distance is preserved in the projection, the corresponding pair is on the bisector (indicated in the figure). If the pair is under the bisector, it represents the projection as an unfolding of the input data manifold.

If the manifold is linear, all points tend to lie on or around (because of noise) of the bisector. Clusters of points on the bisector, but far from the origin, represent large inter-cluster distances, and, hence, reveal the presence of clusters. Figure 21.4 shows this plot for a five-dimensional latent space (this choice is suggested by the previous PCA). The smaller grouping near the origin represents the intra-cluster distances and suggests the idea of one or several hyperplanes as data manifold. This is confirmed by the other groupings at larger distances. They represent the inter-distances and suggest

Fig. 21.3 Variance explained by principal components

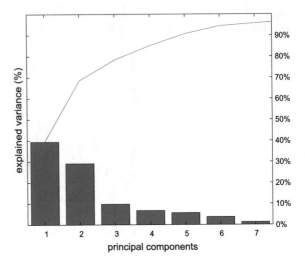

Fig. 21.4 CCA dy-dx diagram

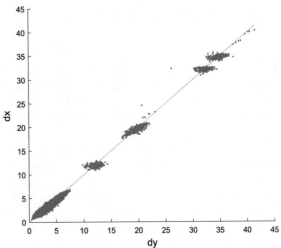

the presence of at least four distinct clusters. The fact that, above all, the groupings of farthest distances have the biggest departure from the bisector, yields the idea of a curvature at large scale. Resuming, the data manifold in the space of the selected genes is composed of several well distinct nearly-flat submanifolds. This confirms the validity of our approach, in the sense that the extracted features discriminate well with regard to tissues.

Biplots [17] are now used in order to understand the reciprocal behavior (in statistical terms) between all tissues and the selected genes. They are a generalization of scatter plots. A biplot allows information on both samples and variables of a data matrix to be displayed graphically. Samples are displayed as points while variables are displayed as vectors. Figure 21.5 shows the biplot over the first three principal

Fig. 21.5 Biplot over the first three principal components

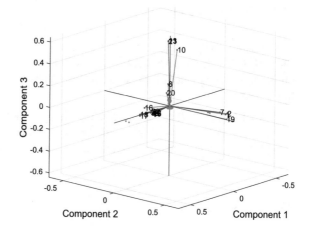

Fig. 21.6 Parallel plot over the resulting gene selection

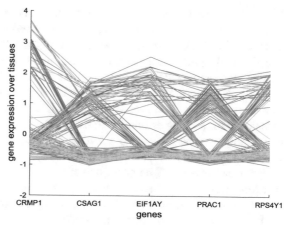

components. With regard to tissues (red points), there are only few data along the first and the second principal component. Instead, the third component has a good discriminant capacity over tissues. With regard to genes (blue vectors), most of them are strongly related to the first or the second principal component. However, five genes (in the figure, represented as 8, 10, 13, 20 and 21) stay along the third axis, thus explaining the variance of tissues along this direction. The biplot shows that a combination of these genes has a bimodal behavior along the third principal component. Figure 21.6 shows a parallel coordinate plot of these five genes. This graph points out relationships between these genes and their bimodal behavior. Blue lines represent worsening cancers, red lines stable cancers and yellow lines regressive cancers, respectively. The most interesting gene is shown in the first vertical axis because it marks a coherent bundle of segments which represents a set of worsening cancers.

21.6 Biological Feedback

The previous analysis shows that the selected genes, whose biological names are CRMP1, CSAG1, EIF1AY, PRAC1 and RPS4Y1 have a high discriminant power. From a biological point of view, some of these genes are strongly related with cancer. In particular:

- CRMP1 is supposed to be related to inhibition of metastasis [18];
- CSAG1 is supposed to be related to squamous cell carcinoma [18];
- PRAC1 is supposed to be related to human prostate and colorectal cancer [18].

21.7 Conclusion

Every patient has his own specific cancer. Hence, personalized precision is required to increase the benefit of therapeutic decisions. This requires an analysis of the gene expression in DNA microarrays, which is very difficult because of the high dimensionality of data in presence of a small training set. Therefore, it has been chosen an unsupervised approach for addressing these problems by using parallel coordinates and a scoring algorithm for the evaluation of the gene power of discrimination. This first gene selection allows the use of linear (PCA) and nonlinear (CCA) reduction techniques for a rough estimation of the geometry of the data manifold. These last considerations pave the way to a further selection of genes, whose properties are checked both from a mathematical and biological point of view. Other criteria of selection can be devised, both by using other techniques of thresholding indices and by taking in consideration the relationships between genes already for the first selection. In this case, parallel coordinate plots may be better considered, above all by studying the geometry of the tissue bundles.

References

1. de Bono, J.S., Ashworth, A.: Translating cancer research into targeted therapeutics. Nature **467**, 543–549 (2010)
2. Hidalgo, M., et al.: Patient-derived xenograft models: an emerging platform for translational cancer research. Cancer Discov. **4**, 998–1013 (2014)
3. Tentler, J.J., et al.: Patient-derived tumour xenografts as models for oncology drug development. Nat. Rev. Clin. Oncol. **9**, 338–350 (2012)
4. Byrne, A.T., et al.: Interrogating open issues in cancer precision medicine with patient derived xenografts. Nat. Rev. Cancer (2017). https://doi.org/10.1038/nrc.2016.140
5. Bertotti, A., et al.: A molecularly annotated platform of patient-derived xenografts ('xenopatients') identifies HER2 as an effective therapeutic target in cetuximab-resistant colorectal cancer. Cancer Discov. **1**, 508–523 (2011)
6. Zanella, E.R. et al.: IGF2 is an actionable target that identifies a distinct subpopulation of colorectal cancer patients with marginal response to anti-EGFR therapies. Sci. Transl. Med. **7** (2015)

7. Bertotti, A., et al.: The genomic landscape of response to EGFR blockade in colorectal cancer. Nature **526**, 263–267 (2015)
8. Sartore-Bianchi, A., et al.: Dual-targeted therapy with trastuzumab and lapatinib in treatment-refractory, KRAS codon 12/13 wild-type, HER2-positive metastatic colorectal cancer (HERACLES): a proof-of-concept, multicentre, open-label, phase 2 trial. Lancet Oncol. **17**, 738–746 (2016)
9. Illumina. Array-based gene expression analysis. Data Sheet Gene Expr. (2011). at http://res.illumina.com/documents/products/datasheets/datasheet_gene_exp_analysis.pdf
10. Sokal, R., Michener, C.: A statistical method for evaluating systematic relationships. Univ. Kansas Sci. Bull. **38**, 1409–1438 (1958)
11. Sokal, R.R., Rohlf, F.J.: The comparison of dendrograms by objective methods. Taxon **11**, 33–40
12. Ward, J.H.: Hierarchical grouping to optimize an objective function. J. Am. Stat. Assoc. **58**, 236–244
13. Calinski, T., Harabasz, J.: A dendrite method for cluster analysis. Commun. Stat. **3**(1), 1–27 (1974)
14. Wegman, Edward J.: Hyperdimensional data analysis using parallel coordinates. J. Am. Stat. Assoc. **85**(411), 664–675 (1990)
15. Jolliffe, I.T.: Principal Component Analysis, Springer Series in Statistics, 2nd edn. Springer, NY (2002)
16. Demartines, P., Hérault, J.: Curvilinear component analysis: a self-organizing neural network for nonlinear mapping of data sets. IEEE Trans. Neural Netw. **8**(1), 148–154 (1997)
17. Gower, J.C., Hand, D.J.: Biplots. Chapman & Hall, London, UK (1996)
18. USA National Center for Biotechnology Information at https://www.ncbi.nlm.nih.gov/

Chapter 22
Computer-Assisted Approaches for Uterine Fibroid Segmentation in MRgFUS Treatments: Quantitative Evaluation and Clinical Feasibility Analysis

Leonardo Rundo, Carmelo Militello, Andrea Tangherloni, Giorgio Russo, Roberto Lagalla, Giancarlo Mauri, Maria Carla Gilardi and Salvatore Vitabile

Abstract Nowadays, uterine fibroids can be treated using Magnetic Resonance guided Focused Ultrasound Surgery (MRgFUS), which is a non-invasive therapy exploiting thermal ablation. In order to measure the Non-Perfused Volume (NPV) for treatment response assessment, the ablated fibroid areas (i.e., Region of Treatment, ROT) are manually contoured by a radiologist. The current operator-dependent methodology could affect the subsequent follow-up phases, due to the lack of result repeatability. In addition, this fully manual procedure is time-consuming, considerably increasing execution times. These critical issues can be addressed only by means of accurate and efficient automated Pattern Recognition approaches. In this contribution, we evaluate two computer-assisted segmentation methods, which we have already developed and validated, for uterine fibroid segmentation in MRgFUS treatments. A quantitative comparison on segmentation accuracy, in terms of area-based and distance-based metrics, was performed. The clinical feasibility of these approaches was assessed from physicians' perspective, by proposing an integrated solution.

L. Rundo · A. Tangherloni · G. Mauri
Dipartimento di Informatica, Sistemistica e Comunicazione (DISCo),
Università degli Studi di Milano-Bicocca, Milan, Italy

L. Rundo · C. Militello (✉) · G. Russo · M. C. Gilardi
Istituto di Bioimmagini e Fisiologia Molecolare - Consiglio Nazionale delle Ricerche
(IBFM-CNR), Cefalù, PA, Italy
e-mail: carmelo.militello@ibfm.cnr.it

R. Lagalla · S. Vitabile
Dipartimento di Biopatologia e Biotecnologie Mediche (DIBIMED),
Università degli Studi di Palermo, Palermo, Italy

© Springer International Publishing AG, part of Springer Nature 2019
A. Esposito et al. (eds.), *Quantifying and Processing Biomedical and Behavioral Signals*, Smart Innovation, Systems and Technologies 103,
https://doi.org/10.1007/978-3-319-95095-2_22

Keywords Computer-assisted medical image segmentation
Pattern Recognition · Magnetic Resonance guided Focused Ultrasound Surgery
Uterine fibroids · Non-Perfused Volume assessment · Clinical feasibility

22.1   Introduction

Magnetic Resonance guided Focused Ultrasound Surgery (MRgFUS) represents a
non-invasive surgical technique that exploits thermal ablation principles to treat several cancer and neurological pathologies, carefully preserving neighboring healthy
tissues (Organs at Risk, OARs) [1]. High Intensity Focused Ultrasound (HIFU) waves
are used to rapidly increase the temperature inside the target solid tumors leading the
neoplastic tissue to either apoptosis or coagulative necrosis [2]. Theoretically, the
HIFU technology can be applied in both small and large irregularly shaped tumors in
any anatomical district where the path to the focus is free of bones and air interfaces
[3]. Image guidance is needed to treat these deep-lying tumors, because accurate
HIFU beam positioning and reliable acoustic power delivery are mandatory. Especially, Magnetic Resonance Imaging (MRI) allows for excellent soft-tissue contrast as
well as real-time visualization of the heat distribution for thermal dose calculation on
target areas, thanks to the Proton Resonance Frequency (PRF) shift thermometry [4].

Uterine leiomyomas, more commonly called fibroids or myomas, are benign
clonal tumors growing from the smooth-muscle tissue of the uterus. They are clinically apparent in about 25% of women during their reproductive years [5], and with
recent in vivo acquisition modalities for soft-tissue imaging, especially Ultrasound
(US) and Magnetic Resonance Imaging (MRI), the actual clinical prevalence may be
higher, so representing a relevant public health problem [6]. Most fibroids are asymptomatic, but many women have significant symptoms that could negatively impact on
their life quality, so requiring effective therapies. Generally, these symptoms include
abnormal uterine bleeding, pelvic pressure or pain, and reproductive dysfunction [7].
Depending on the severity of the symptoms, different treatment options for uterine
fibroids are available: traditional surgery (i.e., abdominal/laparoscopic hysterectomy
or myomectomy), mini-invasive treatments (i.e., hysteroscopic myomectomy, uterine
artery embolization), and non-invasive approaches (i.e., MRgFUS, pharmacological
therapy) [8].

In MRgFUS uterine fibroid therapy, MR image analysis is required in all clinical
phases: (i) imaging-aided diagnosis; (ii) treatment planning and real-time temperature monitoring, by detecting the fibroids to be treated (i.e., Regions of Treatment,
ROTs) as well as the OARs near the uterus (i.e., Region of Interest, ROI) to be preserved from the HIFU beam; (iii) patient's follow-up for measuring the Non-Perfused
Volume (NPV), which is the fibroid area actually ablated with HIFU therapy [9].

In this study, we are particularly interested in MRgFUS treatment response assessment. Nowadays, the treatment outcome is evaluated on complete infarction of the
ablated fibroids and normal perfusion of the surrounding myometrium [10]. This
estimation is accomplished on post-treatment T1-weighted (T1w) contrast-enhanced

MRI, wherein the ablated ROT does not uptake the administered Gadolinium-based contrast medium and are imaged as unenhanced hypo-intense with respect to the uterus. In clinical practice, the NPV is measured by means of manual contouring by an experienced radiologist. The current operator-dependent methodology could affect the subsequent follow-up phases, due to the lack of result repeatability. In addition, this fully manual procedure is definitely time-consuming, considerably increasing treatment times. These critical issues can be addressed only by accurate and efficient computer-assisted MR image segmentation approaches based on Pattern Recognition techniques.

This paper evaluates specifically two validated computer-assisted segmentation methods, which we have already presented in [11, 12], for uterine fibroid segmentation in MRgFUS treatments. A quantitative comparison on segmentation accuracy and reliability, in terms of area-based and distance-based metrics, was performed. The clinical feasibility of these approaches was assessed from physicians' perspective, proposing an integrated solution that considers also the workflow in real environments.

The structure of this manuscript is the following: Sect. 22.2 outlines the state of the art methods for MRI uterine fibroid segmentation; Sect. 22.3 describes the analyzed MRI data as well as our computer-assisted uterine fibroid segmentation approaches; Sect. 22.4 reports the experimental results obtained in the segmentation tests; finally, some discussions and conclusive remarks are given in Sect. 22.5.

22.2 Background

In this section, the most representative literature works on uterine fibroid 2D segmentation approaches are outlined.

Guyon et al. in [13] developed a tool called Volume Estimation and Tracking over Time (VETOT) to track the volume of fibroids in patient's follow-up. VETOT provides two different types of Level Set Functions (LSFs) for fibroid segmentation: (i) fast marching level sets, and (ii) geodesic active contours. In [14], a two-step semi-automatic method was described: an initial automatic segmentation, based on LSFs, is followed by an interactive manual refinement (with user feedback). This method was applied on uterine fibroid MR scans acquired prior to MRgFUS treatment execution. The authors of [15] proposed a semi-automatic approach based on LSFs, by combining the fast marching level set and the Laplacian level set methods. However, this approach was not tailored for HIFU treatments. In [16] a two-step method for segmentation of uterine fibroids was developed: (i) a coarse semi-automatic segmentation is performed using the Chan-Vese level set method; (ii) segmentation refinements based on the prior-shape model are applied, by exploiting training data regarding an ellipse model based on Principal Component Analysis (PCA).

Fallahi et al. in [17] proposed an automated approach for fibroid volume evaluation on MR images. By exploiting the method described in [18], the uterus is firstly segmented using the Fuzzy C-Means (FCM) algorithm from the T1w contrast-enhanced MRI series. Some refinement operations are applied, and redundant parts are removed by masking the co-registered T1w MRI series. The fibroids are then segmented by applying first the Modified Possibilistic Fuzzy C-Means (MPFCM) algorithm on registered T2-weighted (T2w) MR images and finally by performing some post-processing operations. Image registration steps are mandatory and are performed using the external software tool Medical Image Processing, Analysis, and Visualization (MIPAV).

In recent years, our research group has already proposed computer-assisted segmentation approaches for uterine fibroids in MRgFUS treatments [11, 12, 19]. The clinical feasibility of these published and validated methodologies will be evaluated and critically discussed in the next sections.

22.3 Patients and Methods

In this section, firstly MRI data concerning the enrolled subjects are reported, and then our previous automated segmentation methods for uterine fibroid segmentation are described.

22.3.1 Patient Dataset Description

The study was carried out on an MRI dataset composed of 18 patients (average age: 42.22 ± 6.29 years) affected by symptomatic uterine fibroids undergone MRgFUS therapy. The total number of the examined fibroids was 27, since some patients presented a pathological scenario with multiple fibroids.

The analyzed images were acquired using a Signa HDxt MRI 1.5T scanner (General Electric Medical Systems, Milwaukee, WI, USA) at two different institutions. Segmentation tests were performed on MR images acquired after the MRgFUS treatment, executed with the ExAblate 2100 (Insightec Ltd., Carmel, Israel) HIFU equipment. The considered MR slices were scanned using the T1w "Fast Spoiled Gradient Echo + Fat Suppression + Contrast mean" ("FSPGR+FS+C") protocol. This MRI protocol is usually employed for NPV assessment [10], since ablated fibroids appear as hypo-intense areas due to low perfusion of the contrast mean. Sagittal MRI sections were processed, in compliance with current clinical practice for therapy response assessment. MRI data are encoded in the 16-bit DICOM format (Fig. 22.1).

Table 22.1 depicts the relevant MRI acquisition parameters, by reporting the range when different values were used. Two instances of input uterine MR images are shown in Fig. 22.2, where the white contour and the white arrows indicate the uterus ROI and the fibroid ROTs, respectively.

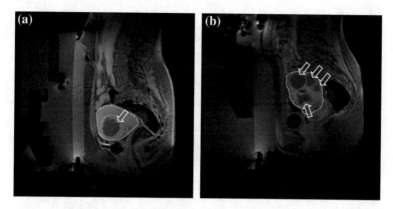

Fig. 22.1 Examples of input T1w "FSPGR+FS+C" MR sagittal slices: **a** large fibroid; **b** multi-fibroid pathological scenario. The uterus ROI is delineated (white contour) and uterine fibroid ROTs are indicated with white arrows

Table 22.1 Acquisition parameters of the MRI uterine fibroid dataset

Acquisition protocol	TR (ms)	TE (ms)	Matrix size (pixels)	Pixel spacing (mm)	Slice thickness (mm)	Spacing between slices (mm)	Number of slices
T1w MRI FSPGR+FS+C	150 ÷ 260	1.392 ÷ 1.544	512 × 512	0.6641 ÷ 0.7031	5.0	6.0	16 ÷ 29

22.3.2 Uterine Fibroid Segmentation Approaches

Uterus and uterine fibroid segmentation on MR images, using Pattern Recognition methods, is a relatively new research field [17], as well as a task of relevant importance in fibroid treatment with the MRgFUS technology. In this section, the challenges and the solutions are thoroughly explained from a clinical usefulness perspective.

Figure 22.2 sketches the overall workflow for uterine fibroid MRgFUS treatment response assessment on post-treatments MRI series.

Uterus ROI Segmentation. Delineating the uterine region is a preliminary step that allows for a robust fibroid detection. This task can be accomplished manually by the user or automatically by means of computational methods. To reduce operator-dependency, our group proposed in [11] a fully automatic approach for uterus segmentation based on the *Fuzzy C-Means (FCM)* algorithm [20, 21], which is an unsupervised Machine Learning technique.

This clustering procedure was performed on gray levels alone and the number of clusters was set to $C = 4$, according to visual inspection (i.e., anatomic properties of the analyzed pelvic images by considering image features) and experimental evidence (by means of segmentation trials). In this automatic ROI detection strategy, we exploited the central position of the uterus in the imaged Field of View (FOV)

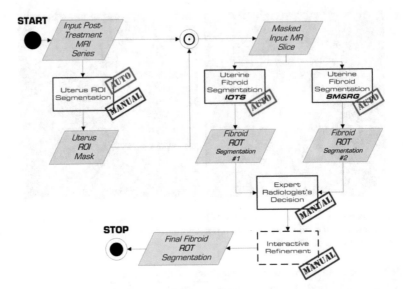

Fig. 22.2 Workflow of uterine fibroid MRgFUS treatment response evaluation. The whole process can be divided into two main stages: (i) uterus ROI segmentation, performed either by an automated approach or manually by an expert physician; (ii) uterine fibroid ROT segmentation, using the proposed approaches based either on the *Iterative Optimal Threshold Selection (IOTS)* algorithm [11] or on combined *Split-and-Merge and Region Growing (SM & RG)* [12]. All the steps involving the radiologist's intervention are also reported, by labeling automatic or manual operations. Solid line boxes are mandatory processes, while dashed line blocks represent interactive steps that are not always required

for ROI cluster selection. As a matter of fact, the ROI is commonly scanned nearby the isocenter of the principal magnetic field to minimize MRI distortions. Thereafter, some post-processing steps were performed in [11] to overcome possible anatomical ambiguities, increasing the robustness of the segmentation method.

A reliable uterus ROI segmentation is clinically useful in MRgFUS treatment planning process, supporting the calibration of the HIFU equipment for OAR protection, as well as in patient's follow-up for an accurate uterus ROI volume measurement [18]. In addition, the segmented uterus ROI may be directly integrated in the subsequent uterine fibroid detection and delineation processing pipeline for MRgFUS treatment response evaluation. In this way, developing operator-independent and reproducible uterine fibroid ROT segmentation approaches will be easier.

Uterine Fibroid ROT Detection and Segmentation. In previous journal articles, we presented two different approaches for uterine fibroid ROT segmentation in MRgFUS treatment evaluation [11, 12]. These adaptive methods relied on a two-phase process, taking advantage of the bimodal histogram of the pre-processed MR image masked with the previously delineated uterus ROI mask. As a matter of fact, uterine fibroid areas undergone thermal ablation reveal a much lower contrast medium absorption than the healthy uterine tissue. In the analyzed post-treatment MR images, ablated fibroids generally appear as homogenous hypo-intense regions with respect to the untreated uterine tissues in T1w contrast-enhanced MRI series. However, an important challenge is represented by the irregular perfusion of the Gadolinium-based contrast medium within fibroid ROTs, due to the heterogeneous dark necrotic tissue caused by the sonication spot distribution for covering the target in HIFU treatment delivery.

The developed methods were designed ad hoc to enhance the current clinical practice in post-operative NPV measurement concerning uterine fibroid MRgFUS treatment outcome assessment:

1. in [11] we presented a method based on the *Iterative Optimal Threshold Selection* (*IOTS*) algorithm [22], by using an efficient implementation that computes image histogram only once [23]. Therefore, a single histogram (i.e., the initial image histogram) is processed during the iterations until convergence. This iterative approach aims at dividing the whole histogram H into two classes H_1 and H_2, separated by an adaptive global threshold $T^{(t)}$, at the t-th iteration. Although *IOTS* works generally well on images characterized by a basically bimodal histogram, some post-processing and mathematical morphological refinements are applied on detected fibroid ROTs to improve segmentation robustness;

2. an advanced direct region detection model for fully automatic fibroid ROT segmentation was proposed in [12]. An incremental procedure was implemented, wherein the results obtained by the *Split-and-Merge* (*SM*) algorithm [24] are employed as multiple seed-region selections by an adaptive *Region Growing* (*RG*) procedure [25]. This combined approach segments multiple fibroids with different pixel intensity, even in the same pelvic MR image. We exploited and improved the semi-automatic multi-seeded region growing approach in [19], which requires interactive seed-point selection. As a result, user input can be omitted, since coarse estimations of the ROTs are automatically detected by means of the region splitting-and-merging procedure according to a Boolean predicate that defines the homogeneity criteria for region segmentation. These seed-regions represent an expressive sample of the ROT to segment with the *RG* procedure [12]. Moreover, an adaptive segmentation threshold, calculated for each ROT in multi-fibroid scenarios using the Otsu's method [26], was introduced in the *RG* similarity conditions.

The two methods may be interchangeably used in the whole workflow defined in Fig. 22.2, representing two different options to effectively support radiologists in accurate NPV evaluation. In both cases, the overall medical image analysis pipeline is unsupervised in all its components, enabling its integration in a comprehensive decision support system that provides clinically useful and quantifiable information (i.e., NPV measurement, ROT shape and location) for personalized medicine.

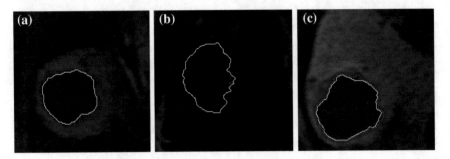

Fig. 22.3 Uterine fibroid ROT segmentation results: segmentation based on *IOTS* (gray contour) and segmentation using *SM & RG* (white contour). (8× zoom factor)

22.4 Result Analysis and Comparison

Area-based and distance-based metrics were considered to evaluate the accuracy of the proposed segmentation methods [27]. The traditional supervised evaluation was employed by comparing the automatically segmented ROIs against the target volume manually contoured by an experienced radiologist (i.e., the gold-standard). Figure 22.3 shows the uterine fibroid ROT delineations, achieved by the *IOTS* and *SM & RG* segmentation methods (over-imposed gray and white contours, respectively).

22.4.1 Evaluation Metrics and Achieved Experimental Results

Area-based metrics measure the pixel-wise similarity of the regions segmented by the proposed method (R_A) against the gold-standard (R_T). Accordingly, the regions containing true positives ($R_{TP} = R_A \cap R_T$), false positives ($R_{FP} = R_A - R_{TP}$), and false negatives ($R_{FN} = R_T - R_{TP}$) are defined with respect to the actual measurement. In our experimental trials, according to the formulations in [11, 12, 21, 27], we used the following measures based on the spatial overlap: Dice similarity index $DSI = 2|R_{TP}|/(|R_A| + |R_T|) \times 100$, Jaccard index $JI = |R_A \cap R_T|/|R_A \cup R_T| \times 100$, Sensitivity $SE = |R_{TP}|/|R_T| \times 100$, and Specificity $SP = (1 - |R_{FP}|/|R_A|) \times 100$.

The spatial distance between the ROT boundaries yielded by the proposed method (defined by the vertices $A = \{\mathbf{a}_i : i = 1, 2, \ldots, K\}$) and the manual contoured boundaries (determined by $T = \{\mathbf{t}_j : j = 1, 2, \ldots, N\}$) was considered to achieve additional insights. Accordingly, the distance between each element of the contour A and the set T must be defined: $d(\mathbf{a}_i, T) = \min_{j \in \{1, \ldots, N\}} \|\mathbf{a}_i - \mathbf{t}_j\|$. The following distance-based measures were used: Mean Absolute Difference $MAD = (1/K) \sum_{i=1}^{K} d(\mathbf{a}_i, T)$, Maximum Difference $MaxD = \max_{i \in \{1, \ldots, K\}} \{d(\mathbf{a}_i, T)\}$,

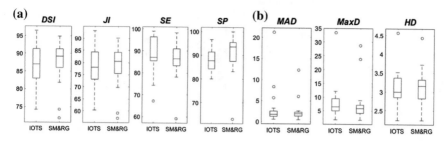

Fig. 22.4 Boxplots of the uterine fibroid ROT segmentation results: **a** area-based metrics; **b** distance-based metrics

and Hausdorff distance $HD = \max\{h(T, A), h(A, T)\}$ (where $h(T, A) = \max_{\mathbf{t} \in T}\{\min_{\mathbf{a} \in A}\{d(\mathbf{t}, \mathbf{a})\}\}$ and $d(\mathbf{t}, \mathbf{a}) = \sqrt{(t_x - a_x)^2 + (t_y - a_y)^2}$).

The values of both area-based and distance-based metrics, achieved by the *IOTS* [11] and *SM & RG* [12] methods on the experimental dataset analyzed in this study, are shown in Table 22.2. To provide a comprehensive graphical statistical summary of the different evaluation metrics, boxplots are illustrated in Fig. 22.4.

DSI and *JI* mean values are approximately equal, proving high segmentation accuracy in both cases. Although *SM & RG* results are characterized by slightly higher standard deviation, the interquartile range in boxplots is more concentrated and two outliers are observed because leaking (i.e., over-estimation of fibroid size) in the case of poor contrast images could occur. *SE* values show that the *IOTS*-based global thresholding solution is often more sensitive than the *SM & RG* approach, especially in little fibroid detected in initial or final ROT slices. Indeed, the region splitting-and-merging procedure could not detect small regions with fuzzy boundaries. However, as visible from the obtained *SP* values and distance-based metrics, the region growing algorithm generally delineates the ROT contours more precisely than the *IOTS* algorithm.

Our approaches have been already compared with similar literature works in the original publications [11, 12]. In both cases, the proposed methods remarkably outperformed the other implemented methods. In addition, the clinical feasibility of many literature approaches is limited. For instance, LSF-based methods are highly sensitive to initial user input and require a labor-intensive parameter setting procedure by clinicians to obtain satisfactory segmentations. Differently to the method proposed in [17, 18], where multispectral MRI sequences and external co-registration operations are mandatory, our methods require T1w contrast-enhanced MRI sequence alone and can be easily integrated in a single software tool.

Table 22.2 Values of area-based and distance-based metrics achieved by the *IOTS* and *SM & RG* methods on the complete MRI dataset composed of 18 patients undergone MRgFUS therapy for uterine fibroid ROT ablation

Method	DSI$_\%$	JI$_\%$	SE$_\%$	SP$_\%$	MAD	MaxD	HD
IOTS [11]	87.25 ± 5.86	78.75 ± 8.74	88.16 ± 8.48	87.88 ± 4.77	3.308 ± 4.863	7.997 ± 6.952	3.100 ± 0.502
SM & RG [12]	87.47 ± 6.29	78.57 ± 9.30	86.07 ± 8.45	90.86 ± 8.23	2.498 ± 2.702	7.890 ± 6.988	3.123 ± 0.494

The results are given as average values ± standard deviations

22.5 Discussion and Conclusions

This paper evaluates two computer-assisted segmentation methods for uterine fibroid segmentation in MRgFUS treatment response assessment, already validated and published in [11, 12] by our research group, considering the clinical feasibility aspects. Our approaches, which exploited Pattern Recognition techniques, were tested on an MRI dataset composed of 18 patients undergone MRgFUS treatment for uterine fibroids, representing an expressive sample of possible clinical scenarios. Segmentation results were quantitatively assessed by means of area-based and distance-based metrics. The evaluation measures achieved by these two methods, namely *IOTS* and *SM & RG* segmentation pipeline, were comparable in terms of delineation accuracy: the former showed higher sensitivity and the latter was characterized by higher specificity. Hence, we can argue that the *IOTS* and *SM & RG* uterine fibroid segmentation approaches have almost complementary properties, allowing for an integrated image processing pipeline by adapting on the different analyzed images. According to Fig. 22.2, a possible clinically feasible solution could be described by the following use case that is easily implementable in the real workflow for uterine fibroid MRgFUS treatment response evaluation:

1. the segmentation results obtained by the two different automatic approaches based on Pattern Recognition algorithms are proposed, by avoiding cognitive overload;
2. the radiologist chooses the best result between the two proposed segmentations, also combining them appropriately in a late integration phase;
3. the user can interactively refine the achieved segmentation by selecting only the correct ROTs (i.e., removing false positives and adding false negatives) as well as by moving the control points that define the computed ROT boundaries.

A volumetric reconstruction of uterus ROI and ablated fibroid ROT could be also useful for radiologists during their own decision-making tasks. Especially, in pathological scenarios with multiple fibroids. However, tridimensional uterine fibroid MR image segmentation approaches are not easy to realize, since fibroids, particularly if they are pedunculated, can be displaced by presenting gaps also in adjacent slices [15]. In such cases, leaking is likely to occur, especially in 3D region growing implementations [28].

In conclusion, the proposed segmentation pipeline is an integrated solution that attempts to solve the clinical problems regarding MRgFUS treatment response evaluation for uterine fibroid, by improving significantly the current operative methodology in terms of segmentation accuracy, result repeatability and execution time. Generally, reliable segmentation results were observed, even when dealing with fibroid ROTs characterized by irregular or inhomogeneous necrotic material, so that operator dependency is reduced. Future works are aimed at the realization of image enhancement methods to enhance the segmentation process for medical images characterized by bimodal histograms, such as the MR images analyzed in this study.

References

1. Jolesz, F.A.: MRI-guided focused ultrasound surgery. Annu. Rev. Med. **60**, 417–430 (2009). https://doi.org/10.1146/annurev.med.60.041707.170303
2. Wu, T., Felmlee, J.P.: A quality control program for MR-guided focused ultrasound ablation therapy. J. Appl. Clin. Med. Phys. **3**(2), 162–167 (2002). https://doi.org/10.1120/jacmp.v3i2.2 584
3. Cline, H.E., Schenck, J.F., Hynynen, K., Watkins, R.D., Souza, S.P., Jolesz, F.A.: MR-guided focused ultrasound surgery. J. Comput. Assist. Tomogr. **16**(6), 956–965 (1992)
4. Agnello, L., Militello, C., Gagliardo, C., Vitabile, S.: Radial basis function interpolation for referenceless thermometry enhancement. In: Advances in Neural Networks—Computational and Theoretical Issues, Smart Innovation, Systems and Technologies, vol. 37, pp. 195–206. Springer, Cham (2015). https://doi.org/10.1007/978-3-319-18164-6_19
5. Buttram Jr., V.C., Reiter, R.C.: Uterine leiomyomata: etiology, symptomatology, and management. Fertil. Steril. **36**(4), 433–445 (1981). https://doi.org/10.1016/S0015-0282(16)45789-4
6. Cramer, S.F., Patel, A.: The frequency of uterine leiomyomas. Am. J. Clin. Pathol. **94**(4), 435–438 (1990). https://doi.org/10.1093/ajcp/94.4.435
7. Stewart, E.A.: Uterine fibroids. Lancet **357**(9252), 293–298 (2001). https://doi.org/10.1016/S 0140-6736(00)03622-9
8. Machtinger, R., Inbar, Y., Cohen-Eylon, S., et al.: MR-guided focus ultrasound (MRgFUS) for symptomatic uterine fibroids: predictors of treatment success. Hum. Reprod. **27**(12), 3425–3431 (2012). https://doi.org/10.1093/humrep/des333
9. Militello, C., Rundo, L., Gilardi, M.C.: Applications of imaging processing to MRgFUS treatment for fibroids: a review. Transl. Cancer Res. **3**(5), 472–482 (2014). https://doi.org/10.397 8/j.issn.2218-676X.2014.09.06
10. Masciocchi, C., Arrigoni, F., Ferrari, F., Giordano, A.V., Iafrate, S., Capretti, I., et al.: Uterine fibroid therapy using interventional radiology mini-invasive treatments: current perspective. Med. Oncol. **34**(4), 52 (2017). https://doi.org/10.1007/s12032-017-0906-5
11. Militello, C., Vitabile, S., Rundo, L., Russo, G., Midiri, M., Gilardi, M.C.: A fully automatic 2D segmentation method for uterine fibroid in MRgFUS treatment evaluation. Comput. Biol. Med. **62**, 277–292 (2015). https://doi.org/10.1016/j.compbiomed.2015.04.030
12. Rundo, L., Militello, C., Vitabile, S., Casarino, C., Russo, G., Midiri, M., Gilardi, M.C.: Combining split-and-merge and multi-seed region growing algorithms for uterine fibroid segmentation in MRgFUS treatments. Med. Biol. Eng. Comput. **54**(7), 1071–1084 (2016). https://do i.org/10.1007/s11517-015-1404-6
13. Guyon, J.P., Foskey, M., Kim, J., Firat, Z., Davis, B., Haneke, K., Aylward, S.R.: VETOT, volume estimation and tracking over time: Framework and validation. In: International Conference on Medical Image Computing and Computer-Assisted Intervention (MICCAI), pp. 142–149. Springer, Berlin (2003). https://doi.org/10.1007/978-3-540-39903-2_18
14. Ben-Zadok, N., Riklin-Raviv, T., Kiryati, N.: Interactive level set segmentation for image-guided therapy. In: Biomed Imaging Nano Macro (ISBI '09), pp. 1079–1082 (2009). https://d oi.org/10.1109/isbi.2009.5193243
15. Yao, J., Chen, D., Lu, W., Premkumar, A.: Uterine fibroid segmentation and volume measurement on MRI. In: Manduca, A., Amini, A.A. (eds.) Medical Imaging 2006: Physiology, Function, and Structure from Medical Images. Proceedings of the SPIE, vol. 6143, pp. 640–649 (2006). https://doi.org/10.1117/12.653856
16. Khotanlou, H., Fallahi, A., Oghabian, M.A., Pooyan, M.: Segmentation of uterine fibroid on MR images based on Chan-Vese level set method and shape prior model. Biomed. Eng. Appl. Basis Commun. **26**(02), 1450030 (2014). https://doi.org/10.4015/S1016237214500306
17. Fallahi, A., Pooyan, M., Khotanlou, H., Hashemi, H., Firouznia, K., Oghabian, M.A.: Uterine fibroid segmentation on multiplan MRI using FCM, MPFCM and morphological operations. In: 2nd International Conference on Computer Engineering and Technology (ICCET), vol. 7, pp. V7-1–V7-5, 16–18 April (2010). https://doi.org/10.1109/iccet.2010.5485920

18. Fallahi, A., Pooyan, M., Hashemi, H., Ghanaati, H., Oghabian, M.A., Khotanlou, H., Shakiba, M., Jalali, A.H., Firouznia, K.: Uterine segmentation and volume measurement in uterine fibroid patients MRI using fuzzy C-mean algorithm and morphological operations. Iran. J. Radiol. 8(3), 150–156 (2011). https://doi.org/10.5812/kmp.iranjradiol.17351065.3142
19. Militello, C., Vitabile, S., Russo, G., Candiano, G., Gagliardo, C., Midiri, M., Gilardi, M.C.: A semi-automatic multi-seed region-growing approach for uterine fibroids segmentation in MRgFUS treatment. In: 7th International Conference on Complex, Intelligent, and Software Intensive Systems (CISIS 2013), pp. 176–182 (2013). https://doi.org/10.1109/cisis.2013.36
20. Bezdek, J.C., Ehrlich, R., Full, W.: FCM: the fuzzy C-means clustering algorithm. Comput. Geosci. 10(2–3), 191–203 (1984). https://doi.org/10.1016/0098-3004(84)90020-7
21. Rundo, L., Militello, C., Russo, G., D'Urso, D., Valastro, L.M., Garufi, A., Mauri, G., Vitabile, S., Gilardi M.C.: Fully automatic multispectral MR image segmentation of prostate gland based on the fuzzy C-means clustering algorithm. In: Multidisciplinary Approaches to Neural Computing, Smart Innovation, Systems and Technologies, vol. 69, pp. 23–37. Springer, Cham (2017). https://doi.org/10.1007/978-3-319-56904-8_3
22. Ridler, T.W., Calvard, S.: Picture thresholding using an iterative selection method. IEEE Trans. Syst. Man Cybern. 8(8), 630–632 (1978). https://doi.org/10.1109/TSMC.1978.4310039
23. Trussell, H.J.: Comments on Picture thresholding using an iterative selection method. IEEE Trans. Syst. Man. Cybern. 9(5), 311 (1979). https://doi.org/10.1109/tsmc.1979.4310204
24. Horowitz, S.L., Pavlidis, T.: Picture segmentation by a tree transversal algorithm. J. ACM 23, 368–388 (1976). https://doi.org/10.1145/321941.321956
25. Adams, R., Bischof, L.: Seeded region growing. IEEE Trans. Pattern Anal. Mach. Intell. 6, 641–647 (1994). https://doi.org/10.1109/34.295913
26. Otsu, N.: A threshold selection method from grey-level histograms. IEEE Trans. Syst. Man Cybern. 9(1), 62–66 (1979). https://doi.org/10.1109/TSMC.1979.4310076
27. Fenster, A., Chiu, B.: Evaluation of segmentation algorithms for medical imaging. In: 27th IEEE Annual International Conference of the Engineering in Medicine and Biology Society (EMBS), pp. 7186–7189 (2005). https://doi.org/10.1109/iembs.2005.1616166
28. Militello, C., Vitabile, S., Rundo, L., Gagliardo, C., & Salerno, S.: An edge-driven 3D region-growing approach for upper airway morphology and volume evaluation in patients with Pierre Robin sequence. Int J Adapt and Innovative Syst. 2(3), 232–253 (2015). https://doi.org/10.15 04/IJAIS.2015.074406.

Chapter 23
Supervised Gene Identification in Colorectal Cancer

P. Barbiero, A. Bertotti, G. Ciravegna, G. Cirrincione, Eros Pasero and E. Piccolo

Abstract Cancer is a large family of genetic diseases that involve abnormal cell growth. Genetic mutations can vary from one patient to another. Therefore, personalized precision is required to increase the reliability of prognostic predictions and the benefit of therapeutic decisions. The most important issues concerning gene analysis are strong noise, high dimensionality and minor differences between observations. Therefore, parallel coordinates have been also used in order to better analyze the data manifold and select the more meaningful genes. Later, it has been chosen to implement a supervised feature selection algorithm in order to work on a subset of features only avoiding the high dimensional problem. Other traditional methods of dimensionality reduction and projection are here used on subset features in order to better analyze the data manifold and select the more meaningful gene. Previous studies show that mutations of genes NRAS, KRAS and BRAF lead to a dramatic decrease in therapeutic effectiveness. The following analysis tries to explore in an

P. Barbiero · G. Ciravegna (✉) · E. Piccolo
DAUIN, Politecnico di Torino, Turin, Italy
e-mail: gabriele.ciravegna@studenti.polito.it

P. Barbiero
e-mail: pietro.barbiero@studenti.polito.it

E. Piccolo
e-mail: elio.piccolo@polito.it

A. Bertotti
Dipartimento di Oncologia, Candiolo Cancer Institute—FPO, IRCCS,
Università degli studi di Torino, Turin, Italy
e-mail: andrea.bertotti@ircc.it

G. Cirrincione
Laboratory of LTI, University of Picardie Jules Verne, Amiens, France
e-mail: exin@u-picardie.fr

G. Cirrincione
University of South Pacific, Suva, Fiji

E. Pasero
DET, Politecnico di Torino, Turin, Italy
e-mail: eros.pasero@polito.it

© Springer International Publishing AG, part of Springer Nature 2019
A. Esposito et al. (eds.), *Quantifying and Processing Biomedical and Behavioral Signals*, Smart Innovation, Systems and Technologies 103,
https://doi.org/10.1007/978-3-319-95095-2_23

unconventional way gene expressions over tissues which are wild type regarding to these genes.

Keywords Colorectal cancer · Curvilinear component analysis · Davies-Bouldin index · Lasso regression · Parallel coordinates · Principal component analysis Xenograft

23.1 Introduction

Medical treatment of cancer is an extremely complex problem, due to intra- and inter-tumour heterogeneity. Therefore personalized or precision medicine approaches are required to increase the reliability of prognostic predictions and the efficacy of therapies [1]. Recently, new powerful tools have been developed for biomarker discovery and drug development in oncology, which rely on a technology called Patient-Derived Xenografts (PDXs) [2–4]. PDXs are obtained by serially propagating surgically-derived tumour specimens in immunocompromised mice. Through this, cancer cells remain viable ex-vivo and retain the typical characteristics of different tumours from different patients. Hence, they can effectively recapitulate the intra- and inter-tumour heterogeneity that is found in real patients. Based on this idea, the PDX technology has been leveraged to conduct large-scale preclinical analyses to identify reliable correlations between genetic or functional traits and sensitivity to anti-cancer drugs. In this context, we have been collecting metastatic colorectal cancer (mCRC) for the last ten years, and we generated the largest PDX biobank available worldwide in an academic environment. Such collection has been widely characterized at the molecular level [5–7] and has been exploited to identify clinically relevant biomarkers [8] for prediction of therapeutic efficacy. Here we propose to exploit available transcriptional data obtained from mCRC PDXs through the Illumina beadarray technology [9] to identify and validate novel predictive algorithms to instruct therapeutic decisions.

23.2 Dataset Analysis and Pre-processing

The dataset stemming from the DNA microarrays is composed of the expression of 15,396 genes in 203 Colorectal Cancer (CRC) murine tissues. For each tissue two additional quantities are available: a discrete variable describing the cancer response to drugs, whose values are chosen as +1 (regressive cancer), 0 (stable cancer) and −1 (worsening cancer); a second continuous variable representing the cancer response to drugs after three weeks, estimated as the difference in size of the tumor.

Data are preprocessed by the z-score technique in order to work on the same range. The analysis has been done by considering the genes as variables. This is a very challenging problem because of analyzing very high dimensional data by means of a small training set. The only possible way to overcome this difficulty is the dimensionality reduction, even if it is data-driven too. As a consequence, the Principal Component Analysis (PCA in an under-determined framework) has been performed both for searching a first rough estimation of the intrinsic dimensionality of data and, above all, to test the nonlinearity of the problem at hand. However, it results that at least 100 principal components are needed in order to explain the 90% of the data variance. This number has the same order of magnitude of the training set size and is a consequence of the high dimensionality and, probably, of the fact that the manifold is nonlinear. Other tools are needed in order to check if the manifold is linear or not. Another possibility is the automatic inspection of the high dimensionality by an explanatory tool. At this purpose, the method of Parallel Plots [10] has been used.

23.3   Parallel Plots

Parallel plots are a powerful way of visualizing high-dimensional data. A point in n-dimensional space is represented as a polyline with vertices on equally spaced parallel axes each one representing a feature.

These plots are used in order to understand in depth the gene capacity of discrimination, by visualizing the average distribution of the murine tissue along all the dimensions (genes) represented as parallel vertical axes. Figure 23.1 only shows the average value for each cancer subtypes together with its 40 and 60% quantiles, in order to easily visualize data distribution. In this way, it is much easier to distinguish meaningful genes. As an example, gene 6 shows a significant distance between different clusterings, suggesting that this gene can be used as marker of CRC subtypes.

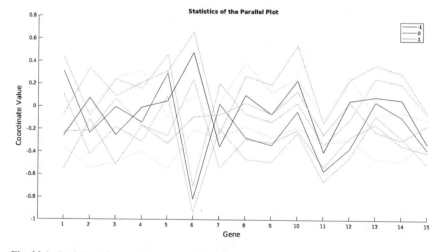

Fig. 23.1 Statistical cluster values in parallel plots

23.4   Feature Selection

A manual feature selection is unfeasible due to the huge amount of features (15,396) and to the difficulty to ascertain the colored bundle groupings. In order to circumvent this problem, a new algorithm for supervised feature selection, based on the Davies-Bouldin [11] clustering index, has been devised. Each gene has been evaluated in its capability of discriminating tissues with different response to drugs which is indeed the skill in grouping well-separated clusters, with high level of cohesion. Specifically, tissues have been divided, for each gene, into unidimensional clusters, exploiting only the associated label. Then, the resulting cluster quality is estimated. The Davies-Bouldin index is suitable for the case of study, because it considers both inter-cluster distances and intra-cluster distances, estimated according to the Euclidean distance.

A quality threshold has been empirically selected in such a way that only the best genes are chosen. Hence, 19 genes have been retained.

This way of selecting feature is unconventional because it does not calculate directly the correlation between genes and the response to drugs, but it still selects the genes that will be more useful and reliable for a classification model based on those genes only.

23.5   Manifold Analysis by PCA and CCA

A subspace composed of the genes selected by the feature selection has been extracted from the original dataset, creating a matrix composed by 203 tissues and only 19 genes. In order to have a better insight of the data extracted, a PCA analysis has been performed for this reduced database. The inspection of the PCA explained variance suggests the intrinsic dimensionality of the manifold to be 11. This could imply data stay on a 11-dimensional hyperplane.

As a further insight of the manifold has been obtained through a Curvilinear Component Analysis (CCA) [12]. This is a neural nonlinear technique of projection which tries to preserve distances in transforming points from the input space to the reduced output space.

However, CCA is here used not for the exploitation of the projection, but for the information that can be derived from its dy-dx diagram (see Fig. 23.2). This plot represents distances between pairs of points in the input space (the dx value) and in the reduced (latent) space (the dy value) as a pair (dy, dx). If a distance is preserved in the projection, the corresponding pair is on the bisector (indicated in the figure). If the pair is under the bisector, it represents the projection as an unfolding of the input data manifold. If the manifold is linear, all points tend to lie on or around (because of noise) of the bisector. Here CCA performs a projection from a 19-dimensional to a 11-dimensional space. Most pairs stay around the bisector, just confirming the hypothesis of 19-dimensional hyperplane.

Fig. 23.2 dy-dx diagram

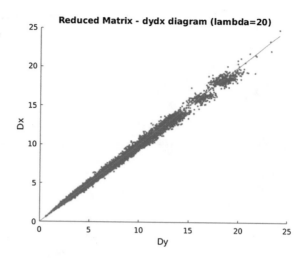

23.6   Lasso Regression

Lasso regression [13] generally targets to improve the prediction accuracy and inter-pretability of the regression model by selecting only a subset of the available variables to use in the final model rather than using all of them. Lasso is able to achieve it by forcing the sum of the absolute value of the regression coefficients to be less than a fixed value, which forces certain coefficients to be set to zero, effectively choosing a simpler model that does not include those coefficients. It requires a regularization λ which controls the trade-off between regression and constraint on the coefficients. Greater values of λ correspond to a lower number of variables inserted in the model.

In the case of study, Lasso regression has been useful to confirm the intrinsic dimensionality of the reduced matrix previously established and, more importantly, to identify the 11 genes. This step has been possible because it is based on the linearity assumption of the reduced database, deduced from the previous manifold analysis. The response variable used previously cannot be exploited here, because it is discrete. Instead, the other variable associated to the tissues could be used (the tumor difference in size after 3 weeks).

Figure 23.3 shows gene coefficients that decrease until zero as λ increases. The value of λ suggested by Lasso (the dotted line in figure) is given by 0.05, because it is the one that guarantees the least MSE. It is important to notice that the number of nonzero LASSO coefficients still present at $\lambda = 0.05$, is 11. This result confirms the previous assumption about the fact that the reduced manifold is a 11-dimensional hyperplane.

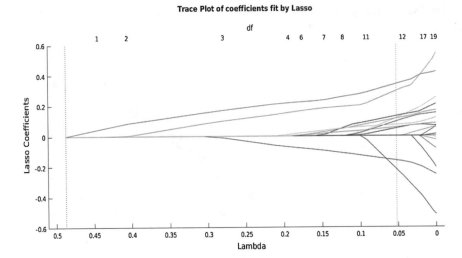

Fig. 23.3 Lasso coefficients as a function of λ

23.7 Bioinformatic Analysis

The 11 genes selected are the most meaningful for predicting the increment or decrement of the cancer volume size after three weeks. Their names are the following: LOC645233, FSCN1, ACSS2, SMAD9, MED1, TMEM118, LOC728505, SF3B4, LOC651316, SERPHIN1, GPR126.

Some of these genes are already well known in the medical literature as correlated with cancer: FSCN1(cell motility [14]), ACSS2 (cancer cell survival) [14], MED1 (gene transcription) [15].

Correlation with cancer of the remaining genes has not been proved yet. Their presence in this work, however, suggests that they should be involved, at least in this particular context of the CRC response to drugs. In fact, it is important to observe that a gene expression may not be relevant for the presence of a tumor, but it may remain important for the survival of tumoral cells. Specifically, average expressions in patients of genes LOC645233 and ACSS2 seems to be in contrast with literature. Nevertheless, results regarding those genes have been published, since they are not an artifact of the analysis but they concern raw data. This analysis proposes a novel approach whose results may be considered as suggestions for further biological research.

23.8   Classification

At last, the expression of the selected 11 genes is used in order to train a classification model. Several models have been tested: the one that shows the best accuracy on the test set is the Support Vector Machine (SVM) [16] with an accuracy of 78%. The model is tested through the hold-out validation with 25% of data randomly put in the test set.

A further attempt to improve the accuracy of the model has been done through the use of a Multilayer Perceptron (MLP) [16]. It is composed of 11 inputs, 20 hyperbolic tangent hidden neurons and 1 output whose activation function is the logistic sigmoid. It is equipped with the cross-entropy error function and the backpropagation learning algorithm is used in order to evaluate the error derivatives for the BFGS training. For the purpose of this analysis two target classes have been selected: the first (1) corresponding to tissues with a regressive or stable response and the second (0) for tissues where the disease has worsened. The robustness of the model is corroborated by both validation and test sets. The accuracy is shown in the Test Confusion Matrix (Fig. 23.4) and is given by 80%.

This result is not only important in itself, but can be considered as a figure of merit for the selected 11 genes: how accurate 11 genes over 15,396 are in modelling the progression of the tumor.

Fig. 23.4 Confusion Matrix for the MLP classifier

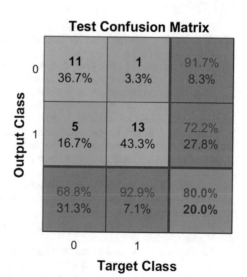

23.9 Conclusion

Every patient has his own specific cancer. Hence, personalized precision is required to increase the benefit of therapeutic decisions. This requires an analysis of the gene expression in DNA microarrays, which is very difficult because of the high dimensionality of data in presence of a small training set. Therefore, it has been chosen a supervised approach for addressing these problems by using parallel coordinates and a scoring algorithm for the evaluation of the gene power of discriminating the tissue response to drugs. This first gene selection allows the use of linear (PCA) and nonlinear (CCA) reduction techniques for a rough estimation of the geometry of the data manifold. These last considerations pave the way to a further selection of genes through the Lasso regression. Eventually, a satisfying classification model has been built on the expression of the chosen genes through the MLP. Other criteria of selection can be devised, both by using other techniques of thresholding indices and by taking in consideration the relationships between genes. In this case, parallel coordinate plots may be better considered, above all by studying the geometry of the tissue bundles. Supervision may help selection from one side, but can also be the goal for the comprehension of the tumor on the other side.

References

1. de Bono, J.S., Ashworth, A.: Translating cancer research into targeted therapeutics. Nature **467**, 543–549 (2010)
2. Hidalgo, M., et al.: Patient-derived Xenograft models: an emerging platform for translational cancer research. Cancer Discov. **4**, 998–1013 (2014)
3. Tentler, J.J., et al.: Patient-derived tumour xenografts as models for oncology drug development. Nat. Rev. Clin. Oncol. **9**, 338–350 (2012)
4. Byrne, A.T., et al.: Interrogating open issues in cancer precision medicine with patient derived xenografts. Nat. Rev. Cancer (2017). https://doi.org/10.1038/nrc.2016.140
5. Bertotti, A., et al.: A molecularly annotated platform of patient- derived xenografts ('xenopatients') identifies HER2 as an effective therapeutic target in cetuximab-resistant colorectal cancer. Cancer Discov. **1**, 508–523 (2011)
6. Zanella, E.R., et al.: IGF2 is an actionable target that identifies a distinct subpopulation of colorectal cancer patients with marginal response to anti-EGFR therapies. Sci. Transl. Med. **7**(272), 272ra12
7. Bertotti, A., et al.: The genomic landscape of response to EGFR blockade in colorectal cancer. Nature **526**, 263–267 (2015)
8. Sartore-Bianchi, A., et al.: Dual-targeted therapy with trastuzumab and lapatinib in treatment-refractory, KRAS codon 12/13 wild-type, HER2-positive metastatic colorectal cancer (HERACLES): a proof-of-concept, multicentre, open-label, phase 2 trial. Lancet Oncol. **17**, 738–746 (2016)
9. Illumina: Array-based gene expression analysis. Data Sheet Gene Expressions at http://res.ill umina.com/documents/products/datasheets/datasheet_gene_exp_analysis.pdf (2011)
10. Wegman, E.J.: Hyperdimensional data analysis using parallel coordinates. J. Am. Stat. Assoc. **85**(411), 664–675 (1990)
11. Davies, D.L., Bouldin, D.W.: A cluster separation measure. IEEE Trans. Pattern Anal. Mach. Intell. 224–227 (1979)

12. Demartines, P., Hérault, J.: Curvilinear component analysis: a self-organizing neural network for nonlinear mapping of data sets. IEEE Trans. Neural Netw. **8**(1), 148–154 (1997)
13. Tibshirani, R.: Regression shrinkage and selection via the lasso. J. Roy. Stat. Soc. (1996)
14. USA National Center for Biotechnology Information at https://www.ncbi.nlm.nih.gov/
15. Human Protein Atlas available from https://www.proteinatlas.org/
16. Christopher, M.B.: Pattern Recognition and Machine Learning. Springer (2016)

Chapter 24
Intelligent Quality Assessment of Geometrical Features for 3D Face Recognition

G. Cirrincione, F. Marcolin, S. Spada and E. Vezzetti

Abstract This paper proposes a methodology to assess the discriminative capabilities of geometrical descriptors referring to the public Bosphorus 3D facial database as testing dataset. The investigated descriptors include histogram versions of Shape Index and Curvedness, Euclidean and geodesic distances between facial soft-tissue landmarks. The discriminability of these features is evaluated through the analysis of single block of features and their meanings with different techniques. Multilayer perceptron neural network methodology is adopted to evaluate the relevance of the features, examined in different test combinations. Principle component analysis (PCA) is applied for dimensionality reduction.

Keywords 3D face recognition · Geometrical descriptors · Dimensionality reduction · Principal component analysis · Neural network

24.1 Introduction

3D face recognition has been deeply investigated in the last decades due to the large number of applications in both security and safety domains, even in real-time scenarios. The third dimension improves accuracy and avoids problems like lighting and make-up variations. In addition, it allows the adoption of geometrical features to study and describe the facial surface.

In this work, the second principal curvature, indicated by k_2, the shape index (S) and the curvedness (C) are used. We rely on the formulations given by Do

F. Marcolin · S. Spada (✉) · E. Vezzetti
DIGEP, Politecnico di Torino, Turin, Italy
e-mail: s219665@studenti.polito.it

G. Cirrincione
Laboratory of LTI, Université de Picardie Jules Verne, Amiens, France
e-mail: exin@u-picardie.fr

G. Cirrincione
University of South Pacific, Suva, Fiji

Carmo [1] for the principal curvatures and by Koenderink and vanDoor for the shape index and curvedness [2]. The shape index describes the shape of the surface. Koenderink and van Doorn proposed a partition of the range $[-1,1]$ in 9 categories, which correspond to 9 different surfaces, ranging from cup to dome/cap, but other representations exist [3, 4]. The partition taken into consideration in this work relies on 7 categories, ranging from cup ($S\in[-1;-0.625]$) to dome ($S\in[0.625;1]$); values of S approximately equal to 0 ($S\in[-0.125;+0.125]$) correspond to saddle-type surfaces. The curvedness is a measure of how highly or gently curved a point is and is defined as the distance from the origin in the (k_1, k_2)-plane. Point-by-point maps of these descriptors of faces we refer to previous work of Vezzetti and Marcolin [4, 5].

The shape index was recently adopted by Quan et al. [6] to build a 3D shape representation scheme for FR relying on the combination of statistical shape modelling and non-rigid deformation matching. The method was tested and compared on the BU-3DFE and GavabDB databases and obtained competitive results even in presence of various expressions. The same descriptor was introduced by Ming [7] for searching nose borders as saddle rut-like shapes in a 3D facial regional segmentation preceding FR relying on a regional and global regression mapping (RGRM) technique. Experiments run on FRGC v2, CASIA, and BU-3DFE databases involving variously "emotioned" faces proved the satisfactory accuracy of the methodology. Ganguly et al. proposed a Range Face image Recognition System (RaFaReS) where the principal curvatures and the curvedness are used as descriptors of the facial surface. The usability of the system was validated on the Frav3D database and results ranged from 89.21 to 94.09% RR depending on the features selected and on the classifier (k-NN and three-layer MLP backpropagation neural network) [8].

In the present work, 7-bins histogram versions of the descriptors are adopted. The choice of 7 bins is given by the definition of the shape index give above. For manageability reasons, the same number of bins is kept for the other descriptors. Besides these descriptors, other features are taken into consideration: the nose volume, Euclidean distances and geodesic distances between typical facial fiducial points, which lie on the skin, called landmarks. The landmarks used here to evaluate these distances are: OE-outer eyebrow, IE-inner eyebrow, EX-exocanthion, EN-endocanthion, N-nasion, AL-alare, PRN-pronasal, SN-subnasal. Except for the eyebrow points, which are not considered real soft-tissue landmarks, their morphometric definitions are provided by Swennen et al. [9].

This paper proposes a methodology to analyze the geometric descriptors and assess their discriminative capabilities. These descriptors are divided in four classes: Euclidean, curvature, shape index and geodesic. Section 24.2 classifies the whole database by using all the descriptors. Section 24.3 introduces the methodology and analyzes the descriptors.

24.2   The 3D Face Database and Its Classification

The dataset is composed of 211 geometrical descriptors (grouped in 4 classes) of 741 faces from 105 subjects. The associated matrix is 741×211. It is statistically normalized by columns (z-score) for having the same range for each descriptor.

The 3D face recognition (classification) is performed by using a Multilayer Perceptron (MLP) [10]. It has a unique hyperbolic tangent hidden layer and a softmax output activation function. The error cost is given by the cross-entropy error. Consequently, the networks outputs the conditional probabilities of class membership.

Training is performed by SCG, which makes use of the error derivatives estimated by the backpropagation algorithm. The network is validated by dividing the dataset in 636 samples for the training set and the remaining ones (one face per person) for validation. All descriptors are taken in account. Hence MLP has 191 inputs. The number of hidden neurons for the best bias variance trade-off is 300 for the 90.04% classification rate in validation. Figure 24.1 represents the confusion matrix as a heat map, just showing that is very sparse. However, the small size of the training set may imply the hidden neurons work as templates for the faces. Therefore, this result is not enough satisfactory.

The fact of using all features, with the goal of exploiting their redundancy, yields a maximum achievable accuracy for the classification. The analysis which follows does not try to improve on it. Instead, by deleting groups of descriptors, uses the classification rate of success in order to detect their relative importance. In this sense, the neural approach is here considered as a probe of the sensitivity of the classification to the choice of features.

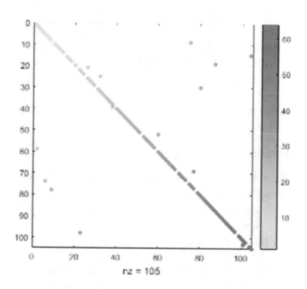

Fig. 24.1 The confusion matrix from MLP as a colored heat map

24.3 Analysis of Geometrical Descriptors

The methodology for the analysis of the geometric descriptors has a two-fold aspect. It does not only consider the advantages and limits for each descriptor in itself and with regard to the other ones, but also makes use of statistical and neural techniques in an unconventional way, i.e. as probes driven by data, for assessing the conclusions.

The statistical tool used here are the following:

- The Principal Component Analysis (PCA), here estimated by solving the SVD problem. Basically, the first principal component (PC) describes the average features of a face, while the subsequent PCs are related to the variance, and so are more important for face recognition.
- The k-means algorithm [11] for clustering.
- The biplots [12] for high dimensional data visualization. They are a generalization of scatter plots. A biplot allows information on both samples and variables of a data matrix to be displayed graphically. Samples are dis-played as points while variables are displayed as vectors.

The neural techniques are given by:

- The curvilinear component analysis (CCA [13]). It is a self-organizing neural network for dimensionality reduction whose neurons are equipped with two weights, the first for the input vector quantization and the second for the projection, which is performed by preserving as many distances as possible in the input space (only distances below a certain threshold λ have to be in-variant in the projection). In particular, the associated dy-dx diagram is important for the following study. This plot represents distances between pairs of points in the input space (the dx value) and in the reduced (latent) space (the dy value) as a pair (dy, dx). If a distance is preserved in the projection, the corresponding pair is on the bisector. If the pair is under the bisector, it represents the projection as an unfolding of the input data manifold. If the manifold is linear, all points tend to lie on or around (because of noise) of the bisector.
- The MLP, configured as seen before, for outputting the conditional probabilities of class membership [10].

More in detail, the steps followed in this work are:

- Analysis of the data (face) manifold: by means of CCA and PCA, the intrinsic dimensionality and the nonlinearity are evaluated, with the aim of defining the smallest set of independent descriptors, in the sense of minimum valuable information.
- Estimation of the descriptor class importance for classification: MLP is used in order to evaluate the quality of one class in classification, both in using only the class of descriptors (single case) and in using the whole matrix except that one (subtractive case); the results are only qualitative and cannot be exactly complementary.

- Assessment of the ability both in discriminating subjects and in clustering the faces as belonging to the same person: it is achieved by the k-means algorithm and checked in a supervised way by counting the number of correct (recognized) labels for each cluster, under the assumption all faces of the same subject (same labels) belong to the same cluster. However, this is a weak hypothesis, because of the possibly different scenarios (e.g. illumination), noise in data and so on. Indeed, using all the descriptors yields a success rate of only 58.44%. Considering that a good clustering cannot be appreciated only in the single case, only the subtractive case is taken in account.
- Study of the statistical properties (covariance) of the descriptor classes and mutual interaction between faces and descriptors; classes are analyzed by using the whole database and in relation to the face information in a double PCA analysis: one in the face space, the other in the descriptor space by interpreting the corresponding biplot.

24.3.1   Geodesic Distances

These distances are more accurate than the Euclidean distances Indeed, in 3D, the geodesic distance between two points on a surface is computed as the length of the shortest path connecting the two points. As a consequence, true distances between facial soft-tissue landmarks are estimated. Instead, the Euclidean distances do not capture this information because they do not follow the shape of the face.

By using PCA on this group of 22 descriptors, it follows that the percentage of explained variance of the first 10 PCs is 91.16% (see Fig. 24.2 left). This fact suggests an intrinsic dimension of 10. At this purpose, CCA is performed for a dimensionality reduction from 22 to 10 ($\lambda = 20$, 80 epochs). The corresponding dy-dx diagram is plotted in Fig. 24.2 right. It shows that the manifold is linear and confirms the validity of the intrinsic dimensionality estimation.

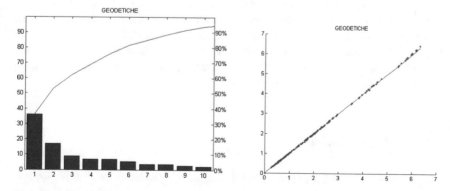

Fig. 24.2 Explained variance by PCA (left) and dy-dx plot by CCA (right)

Fig. 24.3 3D biplot: faces
(red points), blue vectors
(geodesic class), green
vectors (all the other classes)

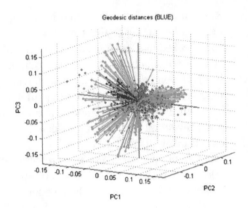

It can be deduced that all faces represented by these descriptors lie on a 10-dimensional hyperplane. This results in a redundancy of the descriptors. Only 10 geodesic distances (or linear combinations of these ones) are needed. Also, faces outside this hyperplane cannot be well represented.

The analysis by MLP yields the following results:

1. Single case: test rate of about 27% with 150 hidden neurons.
2. Subtractive case: test rate of about 89% with 300 hidden neurons.

Considering that the subtractive result is similar to the complete one in Sect. 24.2, it follows that these descriptors are not important in itself, but only in conjunction with other geometric descriptors.

The k-means approach yields a success rate of 56.01% for the subtractive case. This is only slightly lower than in the global case. It means that this class is not necessary for the clustering ability (in the sense that all other descriptors are enough for achieving nearly the same accuracy).

The statistical analysis uses the 3D biplot (see Fig. 24.3). The blue vectors, which represent the geodesic distances, are nearly orthogonal to the plane PC1-PC2, which means they are insensitive to the average values of the faces, but depend on their variance. However, they are clustered, which implies the intrinsic dimension is low. The fact their moduli are large implies their variance is high (they well represent a portion of the descriptor space). With regard to the position of faces (red points), the orthogonality implies the PCA scores are very low, thus confirming the little importance of this class in the classification.

24.3.2 Shape Index

The class of shape indices contains 42 descriptors. By using PCA on this group, it follows that the percentage of explained variance of the first 30 PCs is 96.74% (see Fig. 24.4 left). This fact suggests the importance of most of these features. CCA is

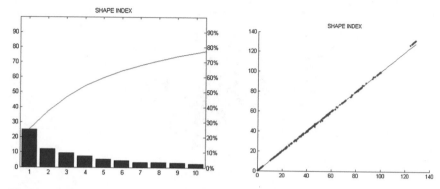

Fig. 24.4 Explained variance by PCA (left) and dy-dx plot by CCA (right)

per-formed for a dimensionality reduction from 42 to 30 ($\lambda = 50$, 120 epochs). The corresponding dy-dx diagram is plotted in Fig. 24.4 right. It shows that the manifold is linear and suggests the idea of a 30-dimensional set of hyperplanes. Indeed, the clusters in the CCA diagram show large distances: they represent intercluster distances, which can be justified either as distant clusters or, more probably, as (face) outliers.

The analysis by MLP yields the following results:

1. Single case: best test rate of about 78.09% with 200 hidden neurons.
2. Subtractive case: best test rate of about 58.02% with 150 hidden neurons.

It can be deduced these indices have a strong impact in face recognition.

The k-means approach yields a success rate of 51.69% for the subtractive case. This is lower than for the geodetic subtractive case and is a confirmation of the better validity of the shape index class.

The statistical analysis uses the 3D biplot (see Fig. 24.5). The blue vectors, which represent the shape indices, are nearly orthogonal to the plane PC1–PC2, which means they are insensitive to the average values of the faces as for the geodetic distances. On the contrary, they spread the PC2–PC3 plane, which implies a good detection of variance of data and confirms the high intrinsic dimensionality. The length of the vectors suggests a high variance for each descriptor (better description of the face).

24.3.3   Indices of Curvature

The curvature class is composed of 84 indices from histograms of the coefficient (42) and the Curvedness (42). The percentage of explained variance (see Fig. 24.6 left) of the first 30 PC components is 98.44% (the first PC is about 50%). It means that a lot of indices are meaningful. This intrinsic dimension is confirmed by CCA

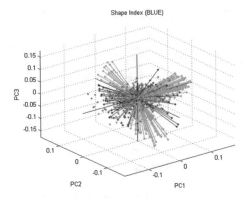

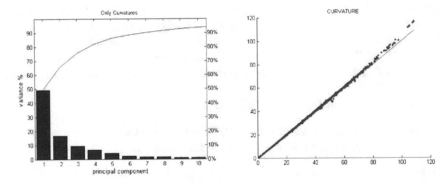

Fig. 24.5 3D biplot: faces (red points), blue vectors (geodesic class), green vectors (all the other classes)

Fig. 24.6 Explained variance by PCA (left) and dy-dx plot by CCA (right)

performed for a dimensionality reduction from 84 to 50 ($\lambda = 50$, 120 epochs), whose dy-dx dia-gram in Fig. 24.6 right shows that the manifold is linear and suggests the idea of a 50-dimensional f hyperplane with some outliers.

MLP yields the following results:

1. Single case: best test rate of about 56.19% with 100 hidden neurons.
2. Subtractive case: best test rate of about 60% with 170 hidden neurons.

It can be deduced these indices have a strong impact as for the shape index, but are slightly less important.

The k-means approach yields a success rate of 58.84% for the subtractive case. This is of the same order of the geodesic subtractive case and proves worse discrimination w.r.t. the shape index.

The statistical analysis by the 3D biplot (see Fig. 24.7) shows the curvature vectors partially spread the PC2–PC3 plane, in a negative correlated way, which implies a good detection of variance, also because of the high score of their projection on data.

Curvatures (BLUE)
0.15
0.1
0.05
0
-0.05
-0.1
-0.15
PC3
0.1
0
-0.1
PC2
-0.1
0
0.1
PC1

Fig. 24.7 3D biplot: faces (red points), blue vectors (curvature class), green vectors (all the other classes)

1. OE - Outer eyebrow
2. IE - Inner eyebrow
3. EX - Exocanthion
4. EN - Endocanthion
5. N - Nasion
6. AL - Alare
7. PRN - Pronasale
8. SN - Subnasale

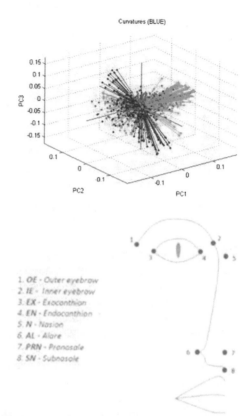

Fig. 24.8 Right facial points used to compute distances. The same are obtained in the left part of the face

The length of the vectors also suggests a high variance for each descriptor (better description of the face). However, some average statistics of faces is captured.

24.3.4 Euclidean Distances

The Euclidean class is composed of 62 distances derived from facial soft-tissue landmarks located around the nose and the eyes (see Fig. 24.8).

The percentage of explained variance (see Fig. 24.9 left) of the first 10 PC components is 93.7%. The intrinsic dimensionality is very low. This is confirmed by CCA for a dimensionality reduction from 62 to 8 ($\lambda = 70$, 120 epochs), whose *dy-dx* diagram in Fig. 24.9 right. It proves the Euclidean manifold is at least an 8-dimensional hyperplane.

However, after a correlation analysis which has selected 8 independent features, it has been proved that 5 of them have a bimodal statistics and are not able to discriminate 100 faces (they always yield the same value). These distances are Pronasal (PN)

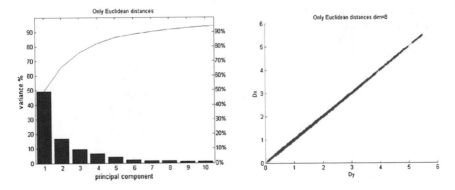

Fig. 24.9 Explained variance by PCA (left) and *dy-dx* plot by CCA (right)

with right Outer eyebrow (OEdx), right Ala of nose (ALA) with right Outer eyebrow, right Exocantion with right Outer eyebrow, left Exocantion with left Outer eyebrow and right Outer eyebrow with left Outer eyebrow. This observation is probably justified by the fact that these facial soft-tissue landmarks are located in positions in which the face movement is difficult. It follows that these distances do not vary for people with different facial expressions. After a detailed examination of these points, the conclusion is that the point 'right Outer eyebrow' (also, the same point in the left position) is a couple of coordinates repeated in all the five distances. This means that: OEdx tracks the change of movement of points PN or ALA when a person modifies his expression. The point Exocantion is located in the inner corner of the eye where there is no possibility of movement. The last distance is between right Outer eyebrow and the analogous in the left position. This distance does not vary for the reason that it is parallel to the frontal anatomical plane and the maximum movement permitted for these points is to shift among the same plane. Resuming, 5 features are certainly unable to classify. There remain only 3 features. MLP confirms the uselessness of this class:

1. Single case: best test rate of about 45.4% with 200 hidden neurons.
2. Subtractive case: best test rate of about 80.95% with 200 hidden neurons.

The k-means algorithm yields a success rate of 70.05% for the subtractive case. It means that the discrimination is better without the Euclidean distances than with the whole database. They worsen the classification.

The 3D biplot (see Fig. 24.10) shows that: the Euclidean distances are clustered around the first component, which represents the average behavior of the faces. Hence, they are not able to well distinguish the images. This clustering also proves the fact they do not discriminate well.

Fig. 24.10 3D biplot: faces
(red points), blue vectors
(Euclidean class), green
vectors (all the other classes)

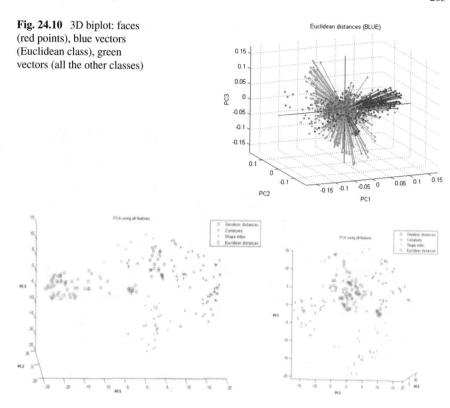

Fig. 24.11 Two views of the scatterplot of the first 3 PCs for all descriptors

24.4 Conclusion

This work analyzes four classes of geometric descriptors for 3D face recognition, by using the public Bosphorus 3D facial database as dataset. At this aim, statistical and neural techniques are employed in an original way. Classes can be ranked according to their behavior in classification. Certainly, the worst features are the Euclidean distances. They do not represent at all the variance in faces. They worsen the face recognition. They are easy to estimate, but not meaning-ful with regard to the true distances in the face, unlike the geodesics, which require a higher computational cost. However, only 10 geodesics are needed. Both curvatures and geodesic descriptors work far better, but they are not enough for capturing the peculiarity of the face. At least 50 curvature descriptors are needed for a good dis-crimination. The best descriptors belong to the shape index class, which requires 30 descriptors. However all these classes but the Euclidean distances have to work in a coordinated way for achieving the best possible result. Figure 24.11 shows the different regions of competence with regard to the first 3 PCs. However, consider that this is only a reduced representation of the feature space (except for the Euclidean class).

Future work will deal with the use of the parallel coordinate plots and other neural networks.

References

1. Do Carmo, M: Differential Geometry of Curves and Surfaces. Englewood Cliffs, New Jersey, Prentice-Hall Inc. (1976)
2. Koenderink, J.J., van Doorn, A.J.: Surface shape and curvature scales. Image Vis. Comput. **10**(8), 557–564 (1992)
3. Dorai, C., Jain, A.K.: COSMOS-A representation scheme for 3D free-form objects. IEEE Trans. Pattern Anal. Mach. Intell. **19**(10), 1115–1130 (1997)
4. Vezzetti, E., Marcolin, F.: Geometrical descriptors for human face morphological analy-sis and recognition. Rob. Auton. Syst. **60**(6), 928–939 (2012)
5. Vezzetti, E, Marcolin, F.: Novel descriptors for geometrical 3D face analysis. Multimedia Tools Appl. **X**, 1–30 (2016)
6. Quan, W., Matuszewski, B.J., Shark, L.K.: Statistical shape modelling for expression-invariant face analysis and recognition. Pattern Anal. Appl. pp. 1–17 (2015)
7. Ming, Y.: Robust regional bounding spherical descriptor for 3D face recognition and emotion analysis. Image Vis. Comput. **35**, 14–22 (2015)
8. Ganguly, S., Bhattacharjee, D., Nasipuri, M.: RaFaReS: range image-based 3D face recognition system by surface descriptor. In: 1st International Conference on Next Generation Computing Technologies (NGCT), pp. 903–908 (2015)
9. Swennen, G.R., Schutyser, F.A., Hausamen, J.E.: Three-dimensional cephalometric: a color atlas and manual. Springer Science & Business Media (2005)
10. Bishop, C.M.: Pattern Recognition and Machine Learning. Springer Science Business Media, LLC © (2006)
11. MacQueen, J.B.: Some methods for classification and analysis of multivariate observations. In: Proceedings of 5-th Berkeley Symposium on Mathematical Statistics and Probability, Berkeley, University of California Press, vol. 1, pp. 281–297 (1967)
12. Gower, J.C., Hand, D.J.: Biplots. Chapman & Hall, London, UK (1996)
13. Demartines, P., Hérault, J.: Curvilinear component analysis: a self-organizing neural network for nonlinear mapping of data sets. IEEE Trans. Neural Netw. **8**(1), 148–154 (1997)

Chapter 25
A Novel Deep Learning Approach in Haematology for Classification of Leucocytes

Vitoantonio Bevilacqua, Antonio Brunetti, Gianpaolo Francesco Trotta, Domenico De Marco, Marco Giuseppe Quercia, Domenico Buongiorno, Alessia D'Introno, Francesco Girardi and Attilio Guarini

Abstract This paper presents a comparison between two different Computer Aided Diagnosis systems for classification of five types of leucocytes located in the tail of a Peripheral Blood Smears: Lymphocytes, Monocytes, Neutrophils, Basophils and Eosinophils. In particular, we have evaluated and compared the performance of a previous feature-based Back Propagation Neural Network classifier with the performance of two novel classifiers both based on Deep Learning using Convolutional Neural Networks introduced in this study. All the classifiers are built considering the same dataset of images acquired in a previous study. The experimental results, reported in terms of accuracy, sensitivity, specificity and precision, show that the different strategies could be compared and discussed from both clinical and technical point of view.

Keywords Computer aided diagnosis · Classification · Artificial neural network
Deep learning · Convolutional neural network · Transfer learning

V. Bevilacqua (✉) · A. Brunetti · D. De Marco · M. G. Quercia · D. Buongiorno
Department of Electrical and Information Engineering (DEI), Polytechnic University of Bari,
Via Orabona 4, 70126 Bari, Italy
e-mail: vitoantonio.bevilacqua@poliba.it

G. F. Trotta
Department of Mechanics, Mathematics and Management (DMMM),
Polytechnic University of Bari, Via Orabona 4, 70126 Bari, Italy

A. D'Introno
Geriatric Medicine-Memory Unit and Rare Disease Centre, University of Bari Aldo Moro, Bari,
Italy

F. Girardi
UVARP ASL Bari, Bari, Italy

A. Guarini
Istituto Tumori Giovanni Paolo II IRCCS, Bari, Italy

25.1 Introduction

In a previous work [1], the authors proposed a system able to count and classify the five types of leucocytes, Lymphocytes, Monocytes, Neutrophils, Basophils and Eosinophils (Fig. 25.1), located in the tail of a Peripheral Blood Smears (PBS) for computing the leucocyte formula, which represents the percentages of the different forms of white blood cells in stained smear.

Image processing and segmentation techniques were used to extract 33 different leucocyte's features (morphological, chromatic and texture-based). Among the 33 previously identified, only 7 features, selected by using the Information Gain Ranking algorithm of Weka platform, were used as input of two different classifiers: a Back Propagation Neural Network (BPNN) and a Decision Tree (DT). Previous comparisons between the two proposed approaches stated that BPNN performed better than the DT on the validation set.

In recent years, Deep Learning architectures spread in the field of Artificial Intelligence, and several innovative Computer Aided Diagnosis (CAD) systems have been introduced, especially in medical field [2]. Specifically, Deep Learning uses Neural Networks, such as Convolutional Neural Networks (CNNs), to learn useful representations of features directly from data and images, thus avoiding some difficult steps of information extraction from data leading to a simpler classification in several fields, e.g. biometrical recognition, cancers detection or anaemia detection [3–5].

Considering Deep Learning architectures based on CNNs, such a kind of systems may be mainly used following two different approaches [2, 6]. The first one consists in the Transfer Learning, whereas the second exploits CNNs as feature extractors for the subsequent classification by a different classifier, such as Artificial Neural Network (ANN) or Support Vector Machine (SVM) [7, 8]. In details, Transfer Learning exploits a pre-trained network and uses it as a starting point to learn a new task; since fine-tuning a network with transfer learning is much faster and easier than training it from scratch, this is a very common approach for the development of classification systems taking images as input. On the other hand, using CNNs as feature extractors could facilitate the creation of input datasets for classifiers taking features as input, e.g. ANNs and SVMs. For this aim, the classification layer and the linked full-connected layer in the CNN must be removed, so that the output of the last considered layer consisted in a set of features describing the input image.

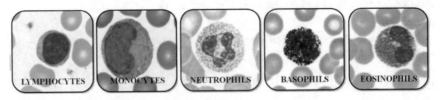

Fig. 25.1 The five types of leucocytes to be classified: (from left to right) Lymphocytes, Monocytes, Neutrophils, Basophils and Eosinophils

Using the MATLAB® NN Toolbox it is possible to deploy CNNs following the two different strategies mentioned before. In this work, the CNN pre-trained model AlexNet [9, 10] was used to design the classifier of leucocytes based on the segmented set of images [1]. Finally, the performances were compared with the ones obtained in the previous work in terms of accuracy, sensitivity, specificity and precision.

The following sections are organized as follow: Sect. 25.2 reports the materials and methods used to acquire and process the images to be classified; Sect. 25.3 reports the classification methods used in this work; Sect. 25.4 shows the results obtained using the implemented classifiers; finally, Sect. 25.5 reports a discussion of the obtained performances.

25.2   Image Processing: Materials and Methods

Blood sampling and preparation for digital acquisition were performed at the National Cancer Research Centre IRCCS "GIOVANNI PAOLO II" (Bari, Italy). The acquired samples were digitalized by means of the microscope D-Sight 200 (www.menarinidi agnostics.it) featuring a 40× optical zoom and the automatic focus was set to reduce the acquisition time. It is noteworthy that thickness irregularity and a non-uniform illumination may affect the digitalization result.

To extract the leucocytes needed to populate the dataset for the Deep Learning architectures, images had to be processed to extract only significant Regions of Interest (ROIs) for the subsequent computation phases. The first step of the proposed approach consisted in the pre-processing of the acquired images. Initially, the entire image was subdivided into sub-images and the segmentation procedure analysed one sub-image at time. This solution was driven by the large size of JPEG2000 files (generally more than 500 MB) that could affect the CAD system performance.

After the previous phase, images were segmented to detect the position of the leucocytes in the obtained sub-images. Thanks to the May-Grunwald-Giemsa staining process, which allows to colour and diversify the blood cells, the nuclei appear to be the most saturated areas. Subsequently, a colour-space conversion from RBG to HSV was performed in order to simplify the nuclei detection, by means of a thresholding process performed considering only the saturation of S channel [11].

Finally, a black and white mask for each sub-image was obtained performing the following operation: (i) connected-component labelling; (ii) identification and removal of blobs with area lower than 20 pixel (which represent noise); (iii) dilatation with rigid circle (diameter equal to 8 pixel) to connect portions of the same nucleus generated by the thresholding process; (iv) identification and removal of blobs with area lower than 400 pixels. The result of each step on a single sub-image is represented in Fig. 25.2.

The obtained mask was used to extract the ROI of each leucocyte (Fig. 25.3), considering the maximum leucocyte's diameter (20 μm) and the coordinates of the nucleus. The ROIs were then used to define the window image as shown in Fig. 25.3; each image was then labelled and the dataset for the subsequent phase of classifica-

(a)　　　　　　　　　　　**(b)**　　　　　　**(c)**

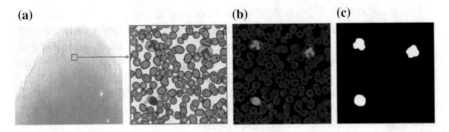

Fig. 25.2 A representation of steps to obtain the nuclei mask [1]; **a** Sub-image extraction. **b** S channel of HSV sub-image. **c** Nuclei mask obtained

Fig. 25.3 Leucocyte's ROI and window extraction in one sub-image [1]

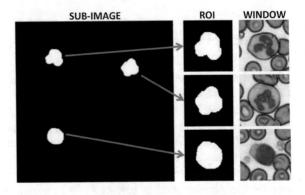

tion was created. Unlike the previous work, since CNNs need images as input, the segmentation steps concerning Plasma detection and Leucocytes edge detection, as well as the features extraction from each ROI Window, were not performed.

25.3 Classification

Deep learning uses CNNs to learn the representation of features directly from images, combining multiple nonlinear processing layers.

Among all the pre-trained models reported in literature, e.g. GoogLeNet [12] and ResNet [13], AlexNet was chosen to perform in both the considered approaches. As reported in [9], AlexNet was trained on approximately 1.2 million images from the ImageNet Dataset (http://image-net.org/index); its model consists in 23 layers and can classify images among 1000 object categories (e.g. keyboard, mouse, coffee mug, pencil, etc.). Thanks to the modularity of the AlexNet architecture (Table 25.1), the layers to use and/or modify change depending on the chosen approach (Transfer Learning or Features Extraction) for classification. Moreover, since the input layer of AlexNet takes only 227 × 227 pixel images, each element of the created dataset was properly scaled to fit the required input size.

Table 25.1 AlexNet architecture

Layer	Transfer function	Description
1	'data'	227 × 227 × 3 images with 'zerocenter' normalization
2	'conv1'	96 11 × 11 × 3 convolutions with stride [4 4] and padding [0 0]
3	'relu1'	ReLU
4	'norm1'	Cross channel normalization with 5 channels per element
5	'pool1'	3 × 3 max pooling with stride [2 2] and padding [0 0]
6	'conv2'	256 5 × 5x48 convolutions with stride [1 1] and padding [2 2]
7	'relu2'	ReLU
8	'norm2'	Cross channel normalization with 5 channels per element
9	'pool2'	3 × 3 max pooling with stride [2 2] and padding [0 0]
10	'conv3'	384 3 × 3 × 256 convolutions with stride [1 1] and padding [1 1]
11	'relu3'	ReLU
12	'conv4'	384 3 × 3 × 192 convolutions with stride [1 1] and padding [1 1]
13	'relu4'	ReLU
14	'conv5'	256 3 × 3 × 192 convolutions with stride [1 1] and padding [1 1]
15	'relu5'	ReLU
16	'pool5'	3 × 3 max pooling with stride [2 2] and padding [0 0]
17	'fc6'	4096 fully connected layer
18	'relu6'	ReLU
19	'drop6'	50% dropout
20	'fc7'	4096 fully connected layer
21	'relu7'	ReLU
22	'drop7'	50% dropout
23	'fc8'	1000 fully connected layer
24	'prob'	Softmax
25	'output'	Cross-entropy with classes

25.3.1 Transfer Learning

Transfer Learning allows to fine-tune a pre-trained Convolutional Neural Network to learn the features on a new collection of images. Following this approach, the last three layers, trained to discriminate among 1000 classes, must be fine-tuned for the new classification task. In other words, it is necessary to replace the last three layers of AlexNet with the new ones recalibrated specifically to perform a five-class discrimination; in particular, they were a fully connected layer, a softmax layer, and a classification output layer.

25.3.2 CNNs as Features Extractors

The second approach consists in Features Extraction, which is the easiest and fastest way to use the representational power of pre-trained deep networks, such as CNNs. The features learned by the convolutional classifier are extracted immediately before the group of layers deputy to the classification.

Considering the AlexNet model, the layer from which the features were extracted is 'fc7' (Table 25.3), which is one of the last full connected layer of the model. In particular, the values considered as features were the activations of the layer, which consists in a 4096-element feature vector. The choice of the extraction layer is a design matter; in fact, it is common use to select, as the designated layer, the FC layer prior to the last one. After the features extraction, an SVM classifier was designed to discriminate among the five classes of leucocytes, considering a one-versus-all approach.

25.4 Results

The Deep Learning classifiers were trained using the same dataset used in [1], but considering image's windows containing a single cell. In detail, the training set, which was composed of 245 leucocytes divided into the five categories, consisted of 110 Neutrophils, 90 Lymphocytes, 30 Monocytes, 11 Eosinophils, and 4 Basophils; the test set, instead, was composed of 1274 leucocytes divided in: 899 Neutrophils, 336 Lymphocytes, 27 Monocytes, 9 Eosinophils, and 3 Basophils.

In the following sections, the classification performances are reported considering the two described approaches; the metrics used for the comparison with the previous work are Accuracy, Sensitivity, Specificity and Precision (Eqs. 25.1–25.4), considering a confusion matrix like the one reported in Table 25.2.

$$Accuracy = \frac{TP + TN}{TP + FP + FN + TN} \tag{25.1}$$

$$Sensitivity = \frac{TP}{TP + FN} \tag{25.2}$$

$$Specificity = \frac{TN}{TN + FP} \tag{25.3}$$

Table 25.2 Confusion matrix

		True condition	
		Positive	Negative
Predicted condition	Positive	*TP*	*FP*
	Negative	*FN*	*TN*

Table 25.3 BPNN confusion matrix

		True condition				
		Basophil	Eosinophil	Lymphocyte	Monocyte	Neutrophil
Predicted condition	Basophil	3	0	0	0	0
	Eosinophil	0	9	2	0	4
	Lymphocyte	0	0	332	2	35
	Monocyte	0	0	2	25	1
	Neutrophil	0	1	0	0	859

Table 25.4 Performance metrics of BPNN classifier

	Sensitivity (%)	Specificity (%)	Precision (%)
Basophils	100	100	100
Eosinophils	90	99.52	60
Lymphocytes	98.80	96.05	89.97
Monocytes	92.59	99.75	89.28
Neutrophils	95.55	99.73	99.88

$$Precision = \frac{TP}{TP + FP} \tag{25.4}$$

25.4.1 BPNN Classifier Results

The classification among Neutrophils, Lymphocytes, Eosinophils, Monocytes and Basophils reached an average accuracy of 96.31%. The Confusion Matrix reporting the BPNN classification results for the best run is reported in Table 25.3. The results of the best run in terms of Sensitivity, Specificity and Precision are reported in Table 25.4.

25.4.2 Transfer Learning Results

The classification among Neutrophils, Lymphocytes, Eosinophils, Monocytes and Basophils reached an average accuracy of 96.38%. The Confusion Matrix reporting the CNN classification results for the best run, considering the Transfer Learning approach, is reported in Table 25.5. The results of the best run in terms of Sensitivity, Specificity and Precision are reported in Table 25.6.

Table 25.5 Transfer learning confusion matrix

		True condition				
		Basophil	Eosinophil	Lymphocyte	Monocyte	Neutrophil
Predicted condition	Basophil	3	0	0	0	0
	Eosinophil	0	6	1	1	1
	Lymphocyte	0	0	330	4	2
	Monocyte	0	0	5	21	1
	Neutrophil	0	1	20	4	874

Table 25.6 Performance metrics of CNN classifier

	Sensitivity (%)	Specificity (%)	Precision (%)
Basophils	100	100	100
Eosinophils	85.71	99.76	66.66
Lymphocytes	92.69	99.34	98.21
Monocytes	70	99.51	77.77
Neutrophils	99.54	93.68	97.21

25.4.3 CNN as Features Extractors Results

The classification among Neutrophils, Lymphocytes, Eosinophils, Monocytes and Basophils reached an average accuracy 95.12%. The Confusion Matrix reporting the results obtained for the best run extracting the features using the CNN and the subsequent classification using the SVM algorithm is reported in Table 25.7. The results of the best run in terms of Sensitivity, Specificity and Precision are reported in Table 25.8.

Table 25.7 SVM classifier confusion matrix

		True condition				
		Basophil	Eosinophil	Lymphocyte	Monocyte	Neutrophil
Predicted condition	Basophil	3	0	0	0	0
	Eosinophil	0	5	1	0	3
	Lymphocyte	0	0	331	2	3
	Monocyte	0	0	5	18	4
	Neutrophil	0	2	15	4	878

Table 25.8 Performance metrics of SVM classifier

	Sensitivity (%)	Specificity (%)	Precision (%)
Basophils	100	100	100
Eosinophils	71.42	99.68	55.55
Lymphocytes	94.03	99.45	98.51
Monocytes	75	99.28	66.66
Neutrophils	98.87	94.55	97.66

25.5 Discussion and Conclusion

White Blood Cells (WBC—or leukocytes) represent the body's main defence against infection. The morphological evaluation of the WBC can help specialists to diagnose hematological pathologies such as leukemia and non-hematological ones such as infectious mononucleosis. The analysis and the counting of blood cells under a microscope can provide useful information on the patient's health status; in particular, the diagnosis of leukemia is possible thanks to the count of white blood cells and morphological analysis of leukocytes. Microscopic observation of blood smears provides important qualitative information on the presence of haematological and extrahemological diseases.

In this work, an innovative CAD system for leucocytes classification based on Deep Learning and Convolutional Neural Network is presented. In particular, two different strategies for the discrimination of the different kinds of cells have been investigated and the performance have been reported in the previous section.

Differently from the previous work, the segmentation steps, including plasma detection and leucocytes edge detection, are not performed because the CNN approach does not require the Features Extraction on the regions of interest.

Moreover, since the average accuracy performance of the CNN Fine-Tuning classifier on the test set is 96.38%, while the CNN-Feature Extraction with SVM classifier is 95.12%, comparing the obtained results with the best performance of the previous BPNN feature-based classifier we can argue that all the approaches seem to be comparable.

In conclusion, the feature-based approach could be more useful in terms of validating previous protocols due to the fact that it highlights the clinical meaning of each feature, while the deep learning approaches could allow a more feasible study due to the fact that could be validated on larger datasets.

References

1. Bevilacqua, V., Buongiorno, D., Carlucci, P., Giglio, F., Tattoli, G., Guarini, A., Sgherza, N., De Tullio, G., Minoia, C., Scattone, A., Simone, G., Girardi, F., Zito, A., Gesualdo, L.: A supervised

CAD to support telemedicine in hematology. In: 2015 International Joint Conference on Neural Networks (IJCNN), pp. 1–7. IEEE (2015)

2. Bevilacqua, V., Altini, D., Bruni, M., Riezzo, M., Brunetti, A., Loconsole, C., Guerriero, A., Trotta, G. F., Fasano, R., Di Pirchio, M., Tartaglia, C., Ventrella, E., Telegrafo, M., Moschetta, M.: A supervised breast lesion images classification from tomosynthesis technique. In: To Appear in 2017 International Conference on Intelligent Computing (ICIC 2017), Lecture Notes in Artificial Intelligence, vol. 9773. Springer (2017)

3. Bevilacqua, V., Cariello, L., Columbo, D., Daleno, D., Fabiano, M.D., Giannini, M., Mastronardi, G., Castellano, M.: Retinal fundus biometric analysis for personal identifications. In: Advanced Intelligent Computing Theories and Applications. With Aspects of Artificial Intelligence. ICIC 2008, Lecture Notes in Computer Science (LNCS), vol. 5227, pp. 1229–1237. Springer (2008)

4. Fakoor, R., Ladhak, F., Nazi, A., Huber, M.: Using deep learning to enhance cancer diagnosis and classification. In: Proceedings of the International Conference on Machine Learning, pp. 1310–1321. ACM (2015)

5. Bevilacqua, V., Dimauro, G., Marino, F., Brunetti, A., Cassano, F., Di Maio, A., Nasca, E., Trotta, G. F., Girardi, F., Ostuni, A., Guarini, A.: A novel approach to evaluate blood parameters using computer vision techniques. In: 2016 IEEE International Symposium on Medical Measurements and Applications (MeMeA), pp. 1–6. IEEE (2016)

6. Erhan, D., Bengio, Y., Courville, A., Manzagol, P. A., Vincent, P., Bengio, S.: Why does unsupervised pre-training help deep learning? J. Mach. Learn. Res. 11(Feb), 625–660 (2010)

7. Bevilacqua, V., Mastronardi, G., Menolascina, F., Pannarale, P., Pedone, A.: A novel multi-objective genetic algorithm approach to artificial neural network topology optimisation: the breast cancer classification problem. In: 2006 International Joint Conference on Neural Networks (IJCNN), pp. 1958–1965. IEEE (2006)

8. Bevilacqua, V., Cassano, F., Mininno, E., Iacca, G.: Optimizing feed-forward neural network topology by multi-objective evolutionary algorithms: a comparative study on biomedical datasets. In: Advances in Artificial Life, Evolutionary Computation and Systems Chemistry, pp. 53–64. Springer (2015)

9. Krizhevsky, A., Sutskever, I., Hinton, G.E.: ImageNet Classification with Deep Convolutional Neural Networks. In Advances in Neural Information Processing Systems, vol. 25, pp. 1097–1105. Curran Associates, Inc. (2012)

10. Vedaldi, A., Lenc, K.: Matconvnet: Convolutional neural networks for matlab. In: Proceedings of the 23rd ACM international conference on Multimedia, pp. 689–692. ACM (2015)

11. Alagappan, M., BanuRekha, B., Arun, R., Kalaikamal, M., Muthukrishnan, S., Ganesh, C.S., Sathishkumar, S.: Extreme learning machine (elm) based automated identification and classification of white blood cells. In: International Conference on Mathematical Modeling and Applied Soft Computing, pp. 846–852 (2012)

12. Szegedy, C., Liu, W., Jia, Y., Sermanet, P., Reed, S., Anguelov, D., Erhan, D., Vanhoucke, V., Rabinovich, A.: Going deeper with convolutions. In: Proceedings of the IEEE Conference on Computer Vision and Pattern Recognition, pp. 1–9. IEEE (2015)

13. He, K., Zhang, X., Ren, S., Sun, J.: Deep residual learning for image recognition. In: Proceedings of the IEEE Conference on Computer Vision and Pattern Recognition, pp. 770–778. IEEE (2016)

Printed in the United States
By Bookmasters